The Philosophical Status of Diagrams

The Philosophical Status of Diagrams

by
Mark Greaves

CSLI
PUBLICATIONS
Center for the Study of
Language and Information
Stanford, California

Center for the Study of Language and Information
Leland Stanford Junior University
Printed in the United States
06 05 04 03 02 5 4 3 2 1

Library of Congress Cataloging-in-Publication Data

Greaves, Mark, 1965–
The philosophical status of diagrams / Mark Greaves.
p. cm. — (CSLI lecture notes ; no. 116)
Includes bibliographical references and index.

ISBN 1-57586-293-X (alk. paper)
ISBN 1-57586-294-8 (pbk. : alk. paper)

1. Logic diagrams. 2. Geometry.
I. Title. II. Series.
BC136 .G74 2001
160–dc21 2001047926
CIP

∞ The acid-free paper used in this book meets the minimum requirements of the American National Standard for Information Sciences—Permanence of Paper for Printed Library Materials, ANSI Z39.48-1984.

CSLI was founded early in 1983 by researchers from Stanford University, SRI International, and Xerox PARC to further research and development of integrated theories of language, information, and computation. CSLI headquarters and CSLI Publications are located on the campus of Stanford University.

CSLI Publications reports new developments in the study of language, information, and computation. In addition to lecture notes, our publications include monographs, working papers, revised dissertations, and conference proceedings. Our aim is to make new results, ideas, and approaches available as quickly as possible. Please visit our web site at
http://cslipublications.stanford.edu/
for comments on this and other titles, as well as for changes and corrections by the author and publisher.

Contents

Acknowledgements

Like many books, this one had an extended gestation period. At various times, it was best described as an essay about the Hyperproof theorem proving environment, as an exploration of a set of fairly narrow logical questions surrounding Hyperproof, and as a treatise on the theory of heterogeneous reasoning. My sincere thanks go to everyone who helped me and supported me during the long period while this project was defining itself.

There are several people and organizations to whom I owe a special debt. It is a pleasure to acknowledge:

John Etchemendy, who showed me how to be a philosopher. I am proud to be his student.

Jon Barwise, John Perry, Johan van Bentham, and the late Wilbur Knorr, who read over the various drafts and helped make it better.

CSLI and the Stanford University Philosophy department, which provided the environment and the infrastructure without which this book could not have be written.

Dave Barker-Plummer, Mike O'Rourke, and Jamie Rucker, who each in their own way taught me how to think.

My mother and father, who always loved me and believed in me, and made sure that I knew it.

Jean, who is simply the best.

Thank you.

1

Introduction

Thinking about reasoning has had a long tradition. Our ability to efficiently construct reliable chains of reasoning is absolutely central to our cognitive abilities. Such a fundamental capability always attracts attention: many noteworthy scholars in history have explicitly addressed themselves to it at one point or another. In Western civilizations, the broad phenomenon of correct reasoning has been seriously and systematically investigated since at least the time of Euclid and Aristotle, and probably before. One result of all this intellectual effort has been the evolution of several disciplines which define precise and distinctive reasoning techniques over particular subject matters. For example, algebra has developed special techniques for reasoning about quantities; geometry has developed special techniques for reasoning about spatial extensions; and classical logic has developed special techniques for reasoning about a certain set of properties which Aristotle believed that all substances must share. In addition, each of these disciplines, as well as many others, have spawned metatheories which describe in a precise way what counts as correct and incorrect reasoning over the individual target domain. These metatheories serve a normative role, by encoding and standardizing the principles of correct reasoning for their individual disciplines. At its most basic level, this book is concerned with the sorts of representations – diagrammatic and sentential – which have been developed for use in an important subset of these theories.

One interesting feature of these accounts of correct reasoning is that, as they have become more precise, the sorts of written representations which they admit as appropriate have converged on a single format. I will call this representational format *sentential*, because the internal structures of these representations exhibit many of the same characteristics as written natural languages. In particular, representations of this kind are typically composed of ordered collections of discrete symbols whose precise geometric properties (except those necessary to determine the symbol ordering) are unrelated to the modeled properties of the domain of reasoning. Sentential-format representations can be opposed to a different possible format for representations, which I will call *diagrammatic*. Diagrammatic representations can be recognized by the extent to which the geometric properties of the components of the representation are relevant to their interpretation, and the ways in which these properties impact the reasoning methods which are licensed by the overall theory. This distinction is not new; several authors have taken note of it and have attempted to make these two roughly defined categories more precise (see, *e.g.*, Shin 1994, Hammer 1995, and Allwein and Barwise 1996). It is also clear that these two formats characterize different endpoints of a spectrum on which actual and potential representational systems can be arrayed – for example, Frege's *Begriffsschrift* notation (which I will discuss in chapter 10) is an example of a blended system which contains elements of both notational styles. At any rate, though, this book will not attempt to contribute to the debate regarding the classification of boundary cases of sentential and diagrammatic representations, as nothing in my argument will hang on fine distinctions in this matter. Rather, I will simply assume: first, that there is such a distinction (rough and theory-laden as it may be); second, that it is possible to reliably recognize many individual cases of the sentential and diagrammatic representational formats; and third, that objects like natural language sentences and predicate calculus formulae fall on the sentential side, and objects like Euler circles and ruler-and-compass geometric graphics fall on the diagrammatic side. Given these assumptions, it becomes possible to consider the historical development of theories of correct reasoning in different disciplines, and explore the reasons for the adoption of one format of representation over another.

This book is concerned with trying to explain a remarkable fact: for a particular set of theories which encode some of our highest standards for correct reasoning – the theories of expression and proof which are operative in geometry and technical logic – the representations which are currently sanctioned are uniformly sentential. This tradition is so pervasive that it is rarely explicitly commented upon; even though textbooks in these subjects may concede that different notations are possible, the overall sentential style is typically presented as if there were no sensible alternative. And, in the occasional cases when this general representational homogeneity is noted, it is usually only to contrast it with the presumed deficiencies of non-sentential representations. Neil Tennant's view is a good example of the received position that diagrammatic representations are formally irrelevant to proper proofs in geometry and formal logic:

> It is now commonplace to observe that the [geometrical] diagram ... is only an heuristic to prompt certain trains of inference; that it is dispensable as a proof-theoretic device; indeed, that it has no proper place in the proof as such. For the proof is a syntactic object consisting only of sentences arranged in a finite and inspectable array. .... Thus the "general triangle" drawn on the page has no genuine role to play in the reasoning. Whatever is true about the correct use of such ploys can be recovered from a proof-theoretic account of various aspects of the inferential patterns among the relevant sentences. The same holds true for the use of Euler-Venn diagrams in monadic predicate logic. .... [This kind of diagram] is as much a chimera as the general triangle. It does no more than recapitulate information already available in the obvious natural deduction of the conclusion from the premises. I have mentioned these two examples of illusory "semantics" or "modeling" because my description of them is widely accepted...[1]

In formal proofs in other areas of mathematics, the dominant view about the role and relevance of diagrams is similar:

> But despite the obvious importance of visual images in human cognitive activities, visual representation remains a second-class citizen in both the theory and practice of mathematics. In particular, we are all taught to look askance at proofs that make crucial use of diagrams, graphs, or other nonlinguistic forms of representation, and we pass on this disdain to our students.[2]

Finally, in spite of the tremendous historical importance of diagrams to the practice of geometry, most non-elementary textbooks routinely start by emphasizing that diagrammatic representations are no longer considered to be legal or meaningful parts of formal geometric proofs, as in:

> We agree to leave "point" and "line" undefined in what follows. We may agree intuitively among ourselves what the standard pictorial representations for these undefined terms will be, but this understanding is no more a part of geometry than a picture is a part of the proof of a theorem.[3]

---

[1] Tennant 1986 pg. 304-5.
[2] Barwise and Etchemendy 1991, pg. 3.
[3] Fishback 1962, pg. 2.

This uniformity of sententially-styled representations in geometry and logic is the central observation which this book seeks to explain.

There are several reasons why this apparent exclusion of diagrammatic representations from current theories of correct reasoning in geometry and logic is particularly striking. The most obvious is that the use of diagrams is ubiquitous in much ordinary nonformal reasoning in these domains. Diagrams are employed as a popular shorthand notation and reasoning substrate in many geometrically-oriented fields, such as mechanics and engineering, as well as in the sorts of logic problems which are popular in puzzle books and standardized examinations. However, beyond this less formal use, diagrams are also routinely employed as part of the standard, approved introduction to more formal reasoning in geometry and logic as well. In logic, training in the technique of Venn diagrams is a widespread and popular part of teaching the basics of the field. And, Fishback's earlier quote notwithstanding, carefully crafted diagrams are an important component of instruction in elementary geometry, and remain pervasive in the texts and journals of advanced geometry, even as their use in formal proofs is officially deprecated. Furthermore, each of these disciplines has had a significant tradition of diagram use in the evolution of its proof standards. Geometry has a long history of officially-accepted deductive techniques which are dependent on exploiting topological features of a drawn diagram, and logic also encompasses several important historical examples of representation and reasoning systems which rely on diagrammatic methods. So, why did these techniques fall out of favor? In the face of clear evidence for the routine and apparently successful use of diagrammatic methods in geometry and logic, what reasons can we point to that would explain their nonformal status in disciplines like these, in which our theories of precise reasoning are the most sophisticated? And specifically, what explains the clearly negative attitude we have observed toward the use of diagrams in geometry and logic today?

These are the general questions which I will answer in this book. In particular, I will show that the answers go beyond the superficial explanation that diagrams are excessively cumbersome to reproduce or communicate orally.[4] While this factor certainly may make the use of diagrams inconvenient, it does not adequately account for the hostility toward diagrams which was noted above. Instead, this book will defend the following broad claim: *The possibility of diagrammatic methods in formal proofs in logic*

---

[4] Evidence that diagrams are cumbersome to produce, if such is necessary, can be found in Hammer 1995. Even with the state-of-the-art production and typesetting equipment at his disposal, Hammer still elected to use linear notation in his chapter on Peirce's existential graphs. See Hammer 1995, pp. 109-110.

and geometry has always been primarily dependent on the characteristics of the metaphysical and ontological theories under which they are carried out.

This claim certainly requires argument. It is not obvious, for example, that the evolution of diagrammatic representations in exact reasoning should not more properly be thought of as driven by different flashes of creativity in different eras: that *e.g.*, the reason that Euler's graphical notation for the syllogism was not invented by Aristotle was simply due to Aristotle's lack of imagination. Another reason often suggested for the marginalization of diagrammatic methods has to do with presumed problems in their application. Certainly, the too-trusting use of diagrammatic methods has been implicated at the heart of some fairly notorious fallacies in geometry, and this has sometimes been taken to imply that the application of these methods has intrinsic problems, especially in contrast to the presumably more well-understood sentential techniques. However, this book will demonstrate that, as a matter of intellectual history, reasons such as these are neither complete nor fundamental. Rather, I will show that the character of the representations used in proofs in logic and geometry can be systematically linked to their philosophical background, and that as this background has evolved, this linkage regulated the different sorts of diagrammatic representations which were developed.

Given that my claim is concerned at root with the historical fortunes of diagrammatic methods of proof, there are two main reasons why examining the evolution of logic and geometry is a natural strategy for uncovering the core issues in such reasoning. First, as I have observed, logic and geometry are examples of disciplines in which our highest current standards for precise reasoning have been achieved. Because of this, the theories which define these standards are some of the most subtle and highly developed of their type, and so any account of the use of diagrams in contemporary proof practice will have to be consistent with these theories. Second, however, both logic and geometry also have a significant tradition of the use of diagrams in proofs, stretching from their disciplinary origins to their current deprecated status today. Geometry in particular has a long history of explicit debate within the discipline about the role of diagrams in axiomatic proof. This feature distinguishes logic and geometry from other disciplines which include formal accounts of correct reasoning where the use of diagrams has never been common (*e.g.*, algebra or number theory), as well as from disciplines like chemistry, which routinely employ structured diagrammatic representations (*e.g.*, reasoning with standardized valence diagrams such as the periodic table), but which do not have a distinct formal account of characteristic disciplinary reasoning.

The argument for my main claim will rely on an analysis of three important common threads, or themes, in the histories of logic and geometry. These are: the role of intuition in the procedures and formalisms of formal proof; the historical contrast between the universalist subject matter of logic and the specific subject matter of geometry; and the way in which both logic and geometry were affected by the changes which swept through mathematics during the nineteenth century. By using these three themes as a lens through which to view the historical development of logic and geometry, I will be able to construct an argument which shows that the history of use of diagrammatic representations in these disciplines cannot be adequately understood without reference to their philosophical background. Let me briefly introduce each of these themes.

The first theme concerns the evolution of current attitudes toward the use of intuition in formal proofs.[5] There are three principal reasons why differing assumptions about the accuracy and dependability of our spatial (or geometric) intuition, in particular, will be important in our investigation of the history of diagrams in logical and geometric proof. First, our ability to manipulate and interpret diagrams in proofs has historically been closely linked to the functioning of our spatial intuition, and so the validity of proofs which rely on diagrams has been thought of as at least partially contingent on the reliability of these intuitions. Second, however, the operation of spatial intuition has also had an important role in guaranteeing that geometric theorems will express truths about the physical world. For most of its history, geometry was accepted as the branch of mathematics whose axioms and inference processes were most clearly reflective of metaphysical truth. This special status was due in large part to the fact that the axioms and construction operations used in geometric proofs were thought to be reliable guides to such truths, primarily because of their direct connection to spatial intuition. Third, in the mathematical flowering of the mid-nineteenth century, both logic and geometry developed very sophisticated views about their ultimate subject matter and the sorts of proof techniques which could be guaranteed to preserve the necessary truths of this subject matter. In both disciplines, these views resulted in an extreme distrust of the reliability of spatial intuition, and a concomitant drive to remove intuitively-based elements from the representation and proof systems that lay at the foundations of each discipline. For these reasons, a careful examination of the role of intuition in the procedures and formalisms of proof will be an important element in the main argument of this book.

---

[5] Of course, the philosophical concept of "intuition" has a complex and subtle history, but in this book I will intend it in a fairly basic sense: as a direct and non-deductive method by which humans can come to know certain classes of truths about the world.

The second theme which I will emphasize concerns the effect of the different subject matters of logic and geometry on the use of diagrams in formal proofs in these fields. Beginning with Aristotle's initial demarcation of the field, logic has historically been charged with capturing and systematizing only our most universal patterns of reasoning, and has therefore shunned any reasoning techniques which are not completely general and uniformly applicable to any subject matter at all. This historical restriction has resulted in more domain-specific methods of reasoning, such as performing a geometric operation on a drawn diagram, only rarely being considered as contained within the scope of purely logical techniques. In contrast, the subject matter of geometry has historically been closely tied to the properties of extension which obtain in a particular (Euclidean) model of the physical world. The ease with which a simple diagram can be used to reason about Euclidean relationships gave rise to a strong diagrammatic tradition in geometry, which persisted even after the invention of analytic geometry. However, the development in the nineteenth century of projective and non-Euclidean geometries resulted in a rapid and radical shift in geometry's officially sanctioned subject matter. By the beginning of the twentieth century, pure geometry had joined logic as a field whose distinctive reasoning techniques were officially viewed as independent from any particular subject matter to which they might be applied. Significantly, this transition in geometry was accompanied by a rejection of diagrammatic methods of proof. Because of this, my argument will closely examine the influence of the perceived subject matter in the evolution of representations and proof standards in logic and geometry.

Finally, this book will concentrate on the importance which the developments in nineteenth-century algebra and analysis held for the evolution of diagrammatic methods in logic and geometry. These developments eventually culminated in the foundational work of David Hilbert. As I will show, the current negative attitude toward diagrams in formal proofs in logic and geometry traces many of its roots to certain features of Hilbert's program for the foundations of mathematics.[6] By examining these roots, I will be able to contribute a new and interesting perspective on Hilbert's philosophy, and at the same time provide an historical backdrop against which to evaluate more recent work in diagrammatic reasoning systems. Although the intellectual

[6] As the twentieth century progressed, several additional factors in the development of modern logic served to reinforce Hilbert's attitude toward diagrammatic proofs. Chief among these are the success of Tarski-style semantics and the rise of philosophical nominalism in the work of Quine and others. However, because these sorts of factors are ultimately dependent on the non-Aristotelian view of logic of which Hilbert was a prime creator, we will not discuss them in this book except in passing.

traditions of logic and geometry were largely separate throughout most of their histories, their intersection in Hilbert's work in the late nineteenth century was a pivotal moment for each field. Briefly, with the work of Boole and Frege on symbolic logic and the work of Klein and Pasch on the hierarchy of systems in geometry, researchers had started to recognize the formal similarities between these two previously separate disciplines. An attempt to construct a unified view of the role of axioms and proof in these theories was inevitable, and it occurred in the work of Hilbert. Hilbert was uniquely qualified to perceive these technical similarities: he was one of the foremost mathematicians of his time, and was deeply interested in foundational issues in both logic and geometry. Over the course of twenty years, his work knit these traditions together into a single theory of axiomatic systems, and established many of the current attitudes regarding representation and reasoning in these disciplines. Not surprisingly, these attitudes are quite similar in many ways, including their rejection of intuition as a tool in formal proofs. This rejection of intuition, as has just been observed, has had serious consequences for the modern acceptability of any sort of diagrammatically-based logical system.

These three themes of intuition, subject matter, and mathematical context in logic and geometry weave together in different ways to help sustain my main argument. Rather than following them directly, however, the organization of this book is primarily temporal. The body consists of two long chronologically-arranged parts: one on the philosophical history of diagram-based methods in geometry, and one on the parallel history in logic. In each of these parts, I will describe the evolution of attitudes toward diagrams in disciplinary representation and reasoning systems, beginning with the Greek origins of the subject, proceeding to the Enlightenment and the intellectual ferment of the eighteenth and nineteenth centuries, and ending with the unifying work of Hilbert in the early 1900s. Even though from a modern point of view the structure of axiomatic systems for geometry and for logic are similar and founded on much of the same concerns, their histories only overlap substantially with the work of Frege and Hilbert at the end of the nineteenth century. Thus, on the surface, these parts may appear relatively independent of each other. However, I will use the three themes I have just described to demonstrate that in their own way, the histories of logic and geometry each trace the working out of different aspects of the central intellectual notions involved in an axiomatic system. And because of this, the perspective afforded by the presence of both histories will allow me to argue that significant changes to the expressive or deductive power of the different theories were invariably accompanied by major changes in the metaphysical or ontological background against which these theories were conceived. In each case, these changes can be shown to support my overall contention that the acceptability of diagram-based systems in logic and geometry has always been primarily dependent on characteristics of the underlying metaphysics and ontology.

Also, in reading this book, it will be important to clearly distinguish two common sorts of questions which have characterized the investigation of precise reasoning in logic and geometry. Questions of the first sort are primarily foundational: one might wonder what features of the world explain the fact that certain patterns of inference are reliable, and certain other patterns are unreliable. Why does the world support successful inference in these disciplines at all, and what accounts for the fact that information flows in the way that it does? Are the reasons for this sort of reliable information flow in the world of the same type as the reasons that explain why, *e.g.*, two pebbles placed next to two pebbles always make four pebbles? The proposed answers to these types of questions typically revolve around the structure and operation of very basic theories of epistemology and metaphysics; for example, Plato's explanation of the possibility of entailments involves an appeal to fundamental relationships between the eternal forms, and a reliance on the power of the dialectical process to expose them. Different accounts of this phenomenon have been given by Aristotle, Hume, Kant, the logical positivists, Hilbert, and many others, and these varying views will be an important underlying issue in this book, especially in the chapters dealing with the metaphysical background to the early theories of Euclid and Aristotle.

Often, however, the subject matters of logic and geometry have been more concerned with questions of a second, much more taxonomic sort: describing, systematizing, and providing a classification for a certain kind of preexisting inferential data, and thereby attempting to supply a method by which one can predict whether or not a particular candidate inference in the discipline is in fact legitimate and reliable. This sort of question is quite different from the foundational question just discussed. Looked at in this way, the task of the logician or geometer does not differ very much from that of any other scientist, except for the extreme generality of her topic. In particular, in order to build a satisfactory theory, she must account for the way that the objects of her theory (*e.g.*, propositions, inference rules, geometry constructions) will relate to the corresponding objects in the data to be explained (*e.g.*, utterances, diagrams, physical measurements of the world). And this task, in turn, requires that a prior determination be made concerning the scope of the actual phenomena to be explained. In short, before significant theoretical work can be done on the structure of these kinds of actual logical or geometric reasoning behavior, agreement must be reached on the proper subject matter of the discipline, and the data against which any candidate theory can be tested. And, unlike the questions of the first sort, making this determination has historically been a matter of intense interest for practicing logicians and geometers. For example, before Aristotle could lay out the theory of the syllogism and its characteristic generality of subject matter, he had to first separate the study of logic from that of rhetoric, and place rhetorical concerns outside the bounds of logic. And, contemporary logicians still by and large ignore issues of, *e.g.*, probabilistic reasoning and reasoning under uncertainty, and justify their exclusion of these phenomena by appealing to the belief that they are not part of the accepted corpus of data which logical theories must explain. Thus, we can see how these sorts of determinations have had dramatic effects on the history of logic and geometry, and because of this we shall have much to say about them in our inquiry.

Finally, before beginning, one caveat is in order. There are definite limits to the degree of certainty which one can attain when trying to explain why a specific intellectual event (in this case, the development of powerful diagrammatically-based inference systems in logic and geometry) did not happen. One can marshal all the relevant evidence, describe the intellectual trends, and speculate about the influence of the zeitgeist and the milieu, but at the end of the day the explainer is always reduced to claiming something like, "the critical ideas occurred to person *X*, and not to person *Y*." Person *Y* did not find the same things obvious that person *X* did. This by itself is not a very convincing explanation. Fortunately, however, it is also not the case that all of intellectual history is simply a more-or-less random progression of ideas which occurred to one person and not another. Philosophical changes in particular rarely happen in a vacuum; they are almost always driven by difficulties with the theories which preceded them. In these cases, an examination of the writings of the various people involved can often help trace how certain ideas evolved and were sharpened as they were transmitted from person to person. Tying an intellectual event to a distinct progression of ideas of this sort results in a much more satisfying explanation for why the event did not occur, and this is the sort of account which I will be attempting here. But my account, like all accounts of this kind, is not able to provide an absolute answer to questions like, "why didn't Aristotle (or Euclid, or Leibniz, or Venn) distinguish cleanly between the syntax and the semantics of the theories they proposed?" The final response to questions like these must be that the relevant ideas just did not occur to them, and have since occurred to us. Further theorizing about why the critical ideas did not emerge yields explanations which should chiefly be viewed as more or less plausible, and only rarely as more or less correct. This unavoidable fact places a limit on the degree of certainty which the account of diagrammatic reasoning offered in this book can achieve.

Part I: Geometry

2

Diagrams for Geometry

Let us turn now to the roots of geometry. Consider the traditional subject matter of plane geometry. This, one would suspect, would be the discipline where diagram-based reasoning techniques would most naturally arise. Diagrams play an indisputable role in the teaching of elementary descriptive geometry: the initial subject matter of mathematical points, lines, and planes has a simple and intuitive graphical representation, and the physical relations between marks that are imposed by a flat drawing surface match well with the abstract relations between objects posited to inhabit the Euclidean plane. Furthermore, diagrams are intimately related to a very important structural issue in classical geometry: many of the common proof strategies are licensed and regulated by spatial properties of the constructed diagrams, and indeed the very definition of what counts as a legitimate proof (as, for example, which cases are distinct and need separate treatment) is often driven by the ability of diagrams to represent various types of geometric information.[1] Nevertheless, professional geometers since the seventeenth century have always sought explicitly to eliminate the role of the diagram, and today the teaching of geometry relegates diagrams to a completely subsidiary, logically derivative role as a mere cognitive aid to the proof. Recall the quote used in the introduction, which comes from Fishback's textbook on projective geometry:

[1] I will discuss this point at greater length below, when I turn to the question of how classical geometric proof practices changed with the introduction of Cartesian representational methods.

> We agree to leave "point" and "line" undefined in what follows. We may agree intuitively among ourselves what the standard pictorial representations for these undefined terms will be, but this understanding is no more a part of geometry than a picture is a part of the proof of a theorem.[2]

This thoroughgoing anti-diagrammatic bias was not Euclid's own; its rise was largely related to the changes in the philosophical foundations of eighteenth and nineteenth-century geometry which we will explore below. However, it is interesting to note that this seems opposite to its counterpart in the history of logic: just as an important part of contemporary research in logic is concerned with an investigation of the pure deductive properties of the specific sentential representation, so one might also expect that research in geometry would at least spawn a branch concerned with the purely deductive properties of diagrammatic representations. But, what happened in logic did not happen in geometry, and indeed contemporary research in geometry and its related fields (differential and projective geometry, topology, etc.) is carried out almost completely with the symbolisms of modern mathematics, and often without the use of any sort of graphical representation of the objects of investigation. In modern axiomatic geometry the elimination of the diagram as a formal reasoning tool has been complete. Mueller, for example, claims:

> One might say that the history of nineteenth-century mathematics is the history of the replacement of geometry by algebra and analysis. There is no geometric truth which does not have a nongeometric representation, a representation which is usually much more compact and useful. Indeed, many mathematicians might prefer to say that traditional or descriptive geometry is simply an interpretation of certain parts of modern algebra.[3]

Given the apparent intuitive importance of diagrams in geometry, then, how can we explain their formal eclipse?

As we will find in our discussion of logic in Part II, many of the traditions of geometry can be traced back to the writings of a single Greek.[4] Indeed, Euclid's dominance of geometry was probably even more complete

---

[2] Fishback 1962, pg. 2.

[3] Mueller 1981, pg. 1.

[4] Apparently, there was very little development of abstract geometry before the Greeks. Kline, for example, claims that, "geometrical thinking in all pre-Greek civilizations was definitely tied to matter. To the Egyptians, for example, a line was no more than either a stretched rope or the edge of a field, and a rectangle was the boundary of a field." This view is shared by Heath. See Kline 1972, pg. 29.

than Aristotle's was of logic. The 13 books of Euclid's *Elements of Geometry*, written within a few decades of the death of Aristotle, defined the subject matter and standards of rigor in geometry for over 2000 years.[5] Until the past century, the system of the *Elements* was the standard way to teach geometry, and it is undisputed that graphical reasoning, although it is often implicit, is a vital part of the traditional Euclidean method. The details of the Euclidean system, however, are largely neutral with regards to a closely related foundational question, one which will be highly relevant to our discussion of the development of diagram use in geometry: namely, the question of what is geometry's ultimate subject matter.

One important answer to this question, held by Aristotle, Hume, Newton, Euler, and more recently revived by Putnam, entails that geometry should be viewed simply as the precise study of the extension of objects in physical space, and therefore (except for the extreme generality of its topic) is not fundamentally different from any other science.[6] In this view, formal geometry is basically just an idealized system for understanding the physical configurations and spatial relations into which bodies enter. The axioms of geometry have the same epistemic and contingent status as the laws of gravity or thermodynamics, and the theorems of geometry are best interpreted as predictions about the actual extensional properties of objects in space. Therefore, geometric theorems would in principle be no different from the predictions of other scientific theories, able to be confirmed or refuted by experimental evidence, and the status of geometric theorems which are not applicable to objects in space would be at best problematic. And, importantly for our purposes, it is a fairly direct consequence of this sort of view that the validity of using diagrams in proofs rests directly on their ability to reliably represent and track real relations obtaining between objects.

Opposed to this sort of empiricism are a number of different views, such as mathematical Platonism and Kantianism, each of which attempts in its own way to secure for geometric theorems the character of necessary truth

[5] The *Elements* are generally dated around 300BC, about 25 years after Aristotle's death. As a matter of historical precedence, Euclid himself cannot be conclusively associated with any of the individual theorems in the *Elements*. Following the scholarship of Heath, Anglin 1994 ascribes the results of Books 1, 2, 6, 7, 8, 9, and 11 to the Pythagoreans, those of Books 3 and 4 to Hippocrates (not the doctor), those of Books 5 and 12 to Eudoxus, and those of Books 10 and 13 to Theaetetus. For a different view on the authorship of the books of the *Elements* and the relationship between Euclid and the philosophers in Plato's Academy, see Knorr 1975. Even granting Euclid's reliance on others for the development of different parts of his geometry, though, his technique of organizing the *Elements* as a cumulative logical sequence from a set of first principles was a first-rank intellectual achievement.

[6] The very word "geometry" is a compound of two sources, which together literally mean the "measurement of the earth."

about spatial extension. Often these views entailed that the subject matter of geometry, rather than being anything empirical, was completely composed of objects created and regulated by the mind. There is also the Hilbert's formalist view, in which geometry does not have a definite subject matter *per se*, but is a reference-free system of axioms and theorems which was originally inspired by observations about the spatial relations that obtain between objects, and which happens to still be able to be used to model certain of them. From these different positions flow different consequences concerning the role of diagrams in geometry, which we will explore. But the central progression is that as empirical attitudes toward geometry diminished in importance, the philosophical basis for allowing diagrams to play a logically significant role in geometric proofs diminished as well. And, as geometric proof techniques and frontier research problems advanced beyond those domains where diagrams were a natural or useful representation, the status of diagrams became genuinely problematic. This observation allows us to suggest a possible answer to the question about the eclipse of diagrams in geometry with which we started this chapter. We will find that despite the ubiquity of diagrams in geometric practice, diagrams actually had a very insecure logical role in proofs because they were linked to a particular, evanescent notion of geometry's subject matter. When that notion of the subject matter became sufficiently removed from human spatial intuitions, diagrams were supplanted by the now-familiar algebraic representations.

The following chapters will be devoted to expanding the themes touched on above. In chapter 3, we will discuss the early history of Euclidean geometry and the roots of its stance toward diagrams. Chapter 4 addresses how the changes in geometric practice brought on by the analytic methods of Descartes affected this position. Chapter 5 deals with the evolution of geometry during the nineteenth century, and is divided into three sections. Section 5.1 considers Poncelet's principle of continuity, and the imaginary geometric elements which this principle licensed. In sections 5.2 and 5.3, we will turn to the effect of non-Euclidean geometry and some of the foundational concerns about mathematics expressed by Frege and Hilbert in the late nineteenth century, and show how these considerations ultimately detached diagrams from the accepted subject matter of geometry and paved the way for the modern view.

3

Euclidean Geometry

It is uncertain which view of his subject matter Euclid himself took.[1] His famous book, the *Elements of Euclid,* itself is not explicitly directed toward to any particular intended domain, *e.g.* the interpretation of surveying data or the exploration of facts about Platonic forms, and larger questions about the metaphysical truth of its geometric statements are never addressed. The Definitions of Book I, with which many of the basic terms of the *Elements* are explained, supply only a set of abstract and physically unfulfillable conditions, the constituent terms of which are not further defined within Euclid's system. For example, we find Euclid defining a *point* as that which has no part (Def. 1), a *line* as breadthless length (Def. 2), and a *surface* as that which has length and breadth only (Def. 5). Interestingly, Euclid does not supply the customary graphical rendering for the fundamental terms in any of the definitions, postulates, or common notions of the *Elements*, nor does he give (except implicitly in the subsequent text) any guidelines for constructing their graphical representatives in a drawn diagram. The lack of an official, textually-supplied interpretation would seem at first to imply that the system of the *Elements* was intended to be applied to anything which can be conceived of in the manner of these definitions. This very modern view, however, neglects the connotations surrounding the terms themselves, the context of Greek geometric research in which the *Elements* is set, and most importantly for our purposes, the role of diagrams in the subsequent structure of the *Elements*. These diagrams, which appear in the various theorems and problems which make up the propositions of the *Elements*, serve to rein-

[1] Heath contents himself with saying that, "[Euclid] may himself have been a Platonist, but this does not follow from the statements of Proclus on the subject. Proclus says namely that he was of the school of Plato and in close touch with that philosophy." Heath 1908 pg. 2. More recent scholarship (cf. Knorr 1975) suggests that the connection between Euclid and Platonism may be even more tenuous than Heath implies.

force the connotations of the initial terms (point, line, plane, figure, boundary, etc.) and ground them in an unmistakably visual context. Further, the particular visual context suggested by the diagrams is not merely an aid for conceptualizing the system of the *Elements*. Through the use of this interpretation of the terms, the geometer is able to assign to their denotata certain other important properties (such as the properties of figures concerning incidence relations, which we will describe below) which are logically essential to Euclid's proofs and which are not described elsewhere in the *Elements*.

In this book, we will assume that the system of the *Elements* was intended to be interpreted as describing a domain of traditional geometric objects whose interrelations respect the constraints of visualizability, consistent with the use of diagrams in the proof practice.[2] Many of Euclid's immediate successors and commentators, such as Proclus, took this domain to be a Platonic one. In this view, the ultimate objects of geometric reasoning (points, lines, triangles, conic sections, etc.) are taken to be real entities existing in the domain of the forms, knowable through the activity of reason, and possessing definite properties that can be discovered. Euclid's method in the *Elements* was designed as a way to systematically reveal these properties from first principles through a progression of theorems and construction problems. In service of this goal, his proof techniques used different types of graphical constructions combined in an intricate way with more familiar sentential-style deductive arguments. Indeed, the first proposition of Book I is a construction problem: "on a given finite straight line to construct an equilateral triangle." In Euclidean practice, the role of the diagram is to symbolize and enforce the constraints of visualizability operating on the geometric figures, and to provide a way to physically represent the workings of these constraints within the context of an individual theorem or construction.

It is a commonplace that the only legal Euclidean diagrams are those which can be constructed with a straightedge and a compass. This is not entirely accurate; Books 11, 12, and 13 of the *Elements* are concerned with solid geometry, and involve the use of conic sections. But even for the initial books, this traditional restriction is not explicitly present in the *Elements*; recall that the *Elements* contain no direct instructions concerning the production of appropriate physical marks. Rather, the limitation to a straightedge and compass is a practical consequence of the constraints on possible geometric figures which are implicit in Euclid's first three postulates. Postulates 1 and 2 license only the production of a straight line between any two points and extending in either direction, and Postulate 3 al-

[2] This view about geometry is in contrast to, for example, Hilbert's. For a differing opinion of the subject matter of Euclidean geometry, see Reed 1995, esp. Chapter 1.

lows the production of a circle with any center and radius.[3] The conventional stipulation that a straightedge and compass are the only legal construction tools for figures in Euclidean plane geometry guarantees that the geometer can not physically draw a diagram whose interpretation would not be consistent with these postulates. Also, relying only on the ability to produce straight lines and circles reinforces the fact (again, guaranteed by the postulates) that the methods used to construct Euclidean figures will be theoretically independent of the necessity of performing precise measurements on those figures. In the following, we will use the term "ruler and compass construction" as a synonym for the sorts of constructions licensed by the postulates of Book 1 of the *Elements*.

The logical force Euclid gained by limiting the possible legal diagrams to those constructible with a ruler and compass functioned to restrict the allowable problems in his domain. At the time Euclid was writing, there was no unanimity about the allowable construction techniques which could be used to define the objects of legitimate geometric study. In particular, the Greeks divided the totality of constructible lines into three broad groups (planar, solid, and linear), depending on the characteristics of their construction.[4] Planar lines, or *loci*, were the simplest, consisting of only straight line segments and circular arcs; solid loci additionally included the parabola, ellipse, and hyperbola (the conic sections); and linear loci were the most sophisticated, encompassing complex curves such as spirals, conchoids, and cissoids. This division gave rise to a corresponding division of geometric problems, so that, for instance, plane problems required only planar loci in their solution. Exhibiting a solution to a given problem classified it as a particular type. For example, in the century before Euclid, the geometer Hippias identified a spiral curve called the quadratrix, and showed that it could be used to trisect an arbitrary angle, thereby showing that angle trisection was a linear problem. However, it was an important (though certainly not absolute) guideline for the ancient geometers that they give solutions to problems using the most restrictive class of loci possible. For example, we find Pappus writing:

---

[3] Two points are worth noting. First, it should be remembered that neither the pure straightedge nor the compass can be used to duplicate a distance; that is, the straightedge cannot be marked, and the compass must "collapse" whenever it leaves the drawing surface (*i.e.*, the compass is a "Euclidean" compass). Propositions 2 and 3 of Book I, however, provide ways to use an unmarked straightedge and a Euclidean compass to transfer a distance. Second, the five postulates of the *Elements*, of course, serve other logical functions besides constraining the type of figures which may be produced in a proof. For example, Postulate 1 and Definition 3 jointly entail that two points are both necessary and sufficient to determine a straight line. For a discussion of these other functions, see Reed 1995 pp. 16-19.

[4] See Heath 1908 pg. 329-331, and Knorr 1986 pg. 341.

> The following sort of thing somehow appears to be no small error to geometers: whenever a planar problem is found by someone via conic or linear [lines], and on the whole whenever [some problem] is solved from a class other than its own...[5]

As Knorr observes, the Greeks had no systematic method of determining that a given problem did *not* belong to a certain class, so even though Hippias had supplied a linear construction for the angle trisection problem, a planar construction was still actively sought. It was only in 1837 that Wantzel proved using the methods of analytic geometry that the trisection of an arbitrary angle was not possible using only constructions allowed under the limitations of plane loci.[6] However, by restricting at the outset the system of the *Elements* to only the geometry of ruler-and-compass constructible figures – that is, to planar loci – Euclid defined his subject matter to be the geometry of figures of the simplest and most fundamental sort.

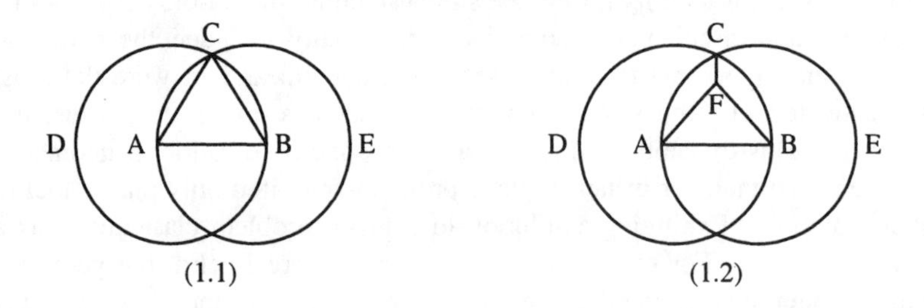

Figure 1: Two Diagrams for *Elements* I-1

Beyond their role in limiting the allowable geometric figures to those constructible with plane loci only, however, lies a more significant consequence of the use of figures constructed with a straightedge and compass. Euclid relies on the intuitive operation of these instruments, or of their idealized counterparts, to ground many of his assertions of existence and inci-

[5] Pappus, *Collection* (IV), I, pp. 370-2. Quoted in Knorr 1986 pg. 345. Knorr points out that this division of problems is in fact more subtle and sophisticated than Pappus suspected.
[6] See Anglin 1994 pg. 204.

dence in a constructed figure.[7] The simplest example of this is in his equilateral triangle construction at *Elements* I 1. The problem of I 1 is to construct an equilateral triangle on a given line AB; Euclid does this as illustrated with diagram 1.1, by drawing circles ABD and BAE, and reasoning that if line segments are drawn to A and B from the point C where the circles cut each other, then the resulting triangle ABC will be equilateral, as in Figure 1.1. Here, Euclid famously (and implicitly) assumes that there will exist a point C at which the circles ABD and BAE will intersect.[8] Further, it has also been observed that Euclid's proof also requires us to assume that two distinct straight lines cannot share a common segment; otherwise, if AC and BC were straight lines sharing segment FC as suggested by Figure 1.2, then the triangle obtained might be ABF, and it would not be equilateral. At the root of each of these unarticulated assumptions are questions concerning the existence of certain geometric objects based on the intersections of other objects, and neither incidence nor the existence of objects arising from it are addressed in Euclid's postulates or common notions.

The modern consensus in response to these observations is to say that they represent a significant flaw in the system of the *Elements*, correctable only via a reformulation along the lines of Hilbert 1899. For our purposes, though, it is interesting to speculate why Euclid, who was so careful in many other ways, would miss this. To me, the most plausible theory is that Euclid relied excessively on the graphical properties of his chosen diagrammatic symbolism, and in this case confused properties of the construction which were explicitly traceable to the operation of the postulates with those which were the result of the construction being carried out with a particular method and medium. Remember that Euclid certainly had some (perhaps hazy) intended domain of figures in mind about which his geometric system was concerned, as well as an intuitive understanding of this intended domain which was logically prior to his formalization. It was surely a fact about the objects populating this domain that, for example, intersecting circles would give rise to an identifiable point. This domain fact not withstanding, though, Euclid's only source of *intrasystem* justification for this sort of fact is the graphical properties of the diagram: specifically, the convention that geo-

[7] The idea of grounding assertions about the existence of geometric figures in the construction techniques which produce them is developed in a much more sophisticated way in Kant's philosophy of geometry. See Friedman 1997.

[8] See Heath 1908 pg. 242. Frege, in §94 of Frege 1884, criticizes *Elements* I 18 along the same lines, claiming that it relies on an unsupported (and presumably diagram-based, though Frege doesn't mention it) inference that a particular point will exist. In I 18, the disputed point is defined by the intersection of a circle and a line, and arises by applying the construction of I 3. Mueller 1981 provides several more examples and strong arguments supporting the pervasiveness of unacknowledged diagram-based reasoning of this sort in the *Elements*.

metric points can be identified where certain constructed lines physically overlay one another.[9] This logical leap from facts about the representation to facts about the domain is seductive, given that one of the advantages of Euclidean diagrammatic representations is that they seem to be necessarily reflective of such existence facts in their classically-understood referents. In the absence of any explicit guidelines about the character and extent of this linkage between the diagram and its domain, however, it also seems clear that the diagram's reflection of these extrasystemic facts should not itself bear any intrasystemic weight.

Even from the modern point of view, though, it is not a very potent criticism of Euclid that these points about the formal status of diagrams seemed to have escaped his attention. In the time of the *Elements*, both logic and mathematics were in their infancy, and these sorts of foundational considerations about the role of the symbolism were not as natural as they are now. Further, it is clear that Euclid definitely did not see the constructed diagram as formally reliable concerning *all* of the possible geometric properties it could represent. We know this because of the fact that his system does not involve unrestricted diagram-based inference; many other kinds of natural diagram-based conclusions (such as those revolving around measurement) were forbidden, and other kinds Euclid appeared to view as problematic. In order to further clarify the logical role of diagrams in Euclid's thought, therefore, it will be instructive to move beyond the details of the construction of I 1 and consider one of these problematic areas in particular: his concept of equality as applied to figures, and the general theory of congruence which is derived from it. In Euclid's system, the basic congruence theorems were founded on observations made after the imagined rigid, or nondeforming, movement of one or more of the geometric figures, and the corresponding operations on the components in the proof's diagram. This general proof technique of employing figure coincidence based on rigid movement has traditionally been called the "method of superposition," and it is reasonably clear that Euclid was troubled by this method. Let us turn to that now.

The method of superposition allows the geometer to assert that two geometric objects are equal, if it can be shown that they would coincide if their corresponding parts were to be "applied" to each other. The legitimacy of superposition in the *Elements* is logically founded on Common Notion 4,

---

[9] Note that this convention could not have been absolute. For example, what if, because of an inaccuracy in drawing, a drawn line physically touches another which is asserted by the problem to be parallel to it? For the Greeks, drawing figures in sand, this sort of inaccuracy must have been common, and its adjudication probably would have rested with the community of geometers.

"things which coincide with one another are equal to one another" (hereafter, CN 4), and Heath makes clear that the Greek term translated here as "coincide" is intended in the physical sense of one thing being physically applied to another.[10] Euclid employed the method of superposition quite sparingly in the *Elements*, initially in his first proof of triangle congruency (the SAS theorem) at I 4. Hilbert included a slightly weaker version of this theorem as an axiom in the geometrical system of Hilbert 1899, and also showed that without superposition and CN 4, I 4 is not deducible in the system of the *Elements*.[11] In order to illustrate Euclid's use of superposition, let us examine Euclid's proof of I 4, which is fairly short and easy to follow, if not completely adherent to modern standards of rigor. Euclid's general strategy in the following is to argue that if triangle ABC is applied to triangle DEF, then the equalities stipulated by the theorem preconditions will force the equal sides to coincide. But then, since these two sides coincide, their endpoints must coincide, and therefore the triangle bases determined by these endpoints must also coincide with each other. Therefore, since the three sides of each triangle coincide with each other, the two triangles themselves must coincide, and so they are equal, and so all of their remaining respective parts must be equal. Here is Euclid's proof:[12]

> **Proposition**: If two triangles have the two sides equal to two sides respectively, and have the angles contained by the equal straight lines equal, they will also have the base equal to the base, the triangle will be equal to the triangle, and the remaining angles will be equal to the remaining angles respectively, namely those which the equal sides subtend.

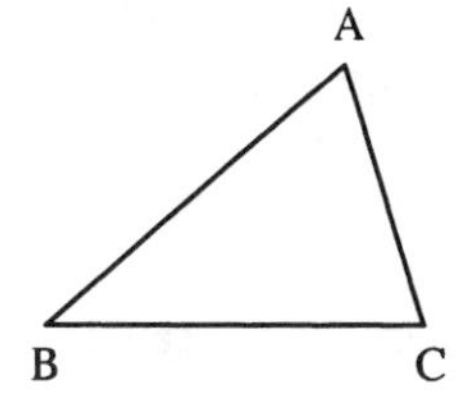

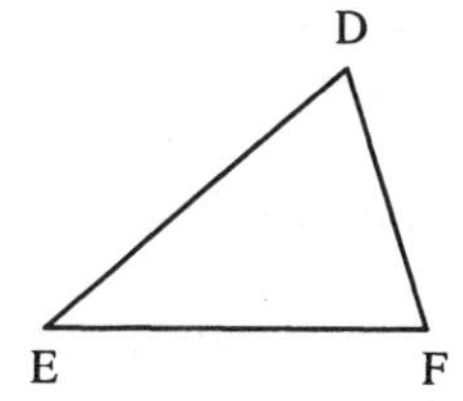

Figure 2: Diagram for *Elements* I-4

[10] Heath 1908, pp. 224-5.

[11] See Hilbert 1899, pp. 39-41. See also Mueller 1981 pg. 23.

[12] This proof is taken from Heath 1908 pg. 247-8.

1. Let ABC, DEF be two triangles having the two sides AB, AC equal to the two sides DE, DF respectively, namely AB to DE and AC to DF, and the angle BAC equal to the angle EDF.

2. I say that the base BC is also equal to the base EF, the triangle ABC will be equal to the triangle DEF, and the remaining angles will be equal to the remaining angles respectively, namely those which the equal sides subtend, that is, the angle ABC to the angle DEF, and the angle ACB to the angle DFE.

3. For, if the triangle ABC be applied to the triangle DEF, and if the point A be placed on the point D and the straight line AB on DE, then the point B will also coincide with E, because AB is equal to DE.

4. Again, AB coinciding with DE, the straight line AC will also coincide with DF, because the angle BAC is equal to the angle EDF; hence the point C will also coincide with the point F, because AC is again equal to DF.

5. But B also coincided with E; hence the base BC will coincide with the base EF, and will be equal to it.

6. Thus the whole triangle ABC will coincide with the whole triangle DEF, and will be equal to it.

7. And the remaining angles will also coincide with the remaining angles and will be equal to them, the angle ABC to the angle DEF, and the angle ACB to the angle DFE. Therefore, [repetition of the statement of the proposition], (being) what it was required to prove.

Here, the use of superposition is signaled with step 3, where triangle ABC is *applied* to triangle DEF. From this, Euclid reasons in steps 3 and 4 that the sides and angles which are specified as equal in the problem statement will coincide. (This inference, from the observation that two elements are equal to the conclusion that they will coincide if applied to one another, is already a bit problematic in Euclid's system, because it rests on the *converse* of CN 4.) Next, Euclid concludes at step 5 that because points B and C will coincide with E and F, respectively, BC will also coincide with EF. This is the critical step, for two reasons. First, step 5 is a deduction from one assertion about coincidence (about points) to another (about segments), and hence it is outside of Euclid's formal system, because the operation of coincidence is never developed in the *Elements* outside of the assertion of CN 4. The validity of step 5 is entirely dependent on preexisting spatial intuitions about the extensional properties of the geometric objects, properties for which a diagram is taken to be a reliable representation. Second, when this deficiency was recognized in antiquity, attempts to patch up

Euclid's proof apparently involved some variant of the following argument, which can now be found as a piece of *scholia* that is normally appended to step 5:

> For if, when B coincides with E and C with F, the base BC does not coincide with the base EF, two straight lines will enclose a space: which is impossible. Therefore the base BC will coincide with EF, and will be equal to it [by Common Notion 4].[13]

But this simply replaces one spatial intuition unsupported in the system of the *Elements* – guaranteeing the coincidence and equality of objects given the coincidence of certain parts – with other, possibly more fundamental, but equally unsupported spatial intuitions concerning the demarcation of a region by lines and the related concepts of interior and exterior. And, although more modern geometers have succeeded in further formalizing these concepts and expressing them rigorously, the point remains that Euclid and his contemporaries were studying a geometry in which equalities derived solely from assumed spatial coincidence played a significant role. The validity of Euclid's actual proof of I 4 depends on the nature of the inferences one can legally make as the result of the method of superposition. These inferences were not captured in his axiom system except via CN 4, and were mediated by the geometer's intuitive understanding of the domain and the understanding imparted by diagrams. Euclid must have sensed this logical problem, as he only uses the method of superposition in a few other proofs, even though its use would have considerably shortened other ones. Both Heath and Mueller, for example, refer to *Elements* I 26 (the ASA triangle congruence theorem) as a theorem whose proof would be shorter if superposition was employed, and Heath dryly recalls Peletarius' 1557 observation that,

> if superposition of lines and figures could be assumed as a method of proof, the whole of geometry would be full of such proofs, that it could equally well have been used in [*Elements*] I 2, 3 ... and that in short it is obvious how far removed the method is from the dignity of geometry.[14]

---

[13] Heath 1908, pp. 248-9. Apparently as a result of the defect of I 4 illustrated here, and the presence of similar defects in other of Euclid's proofs, many ancient manuscripts gave the principle that two straight lines cannot enclose a space as either a sixth postulate or as an additional common notion, and Proclus even supplied a (fallacious) proof of it based on the definition of a circle.

[14] See Heath 1908, pgs. 225 and 249.

Indeed, if we replace Euclid's fourth postulate (that all right angles are equal to each other) by the axiom that figures are invariable when they are moved through space, then we can use superposition to prove the equality of right angles.[15] The fact that Euclid did not do this, and preferred to state the equality of right angles directly, is further evidence that Euclid was uncomfortable with relying on the results of rigid motion, and used it only when he could see no other option.

Considering Euclid's use of the method of superposition allows us to clarify our initial analysis, based on the existence assumptions of I 1, of the interplay between ancient geometric diagrams and the sorts of things they were taken to represent. As we have noted, Euclid presumably did not see himself as reasoning about the physical diagram itself, or some perfected but somehow still physical version of it; rather, the diagram had to be interpreted as a representational device with which to visualize relations which actually took place at the level of the real objects of geometry. This would have been a familiar view to Euclid; we can find Plato himself referring to it in a discussion of geometric practice:

> [geometry students] further make use of the visible forms and talk about them, though they are not thinking of them but of those things of which they are a likeness, pursing their inquiry for the sake of the square as such and the diagonal as such, and not for the sake of the image of it which they draw. And so in all cases. The very things which they mold and draw, which have shadows and images of themselves in water, these things they treat in their turn as only images, but what they really seek is to get sight of those realities which can be seen only by the mind.[16]

Further evidence for the commonality of this view with regards to geometric diagrams can be found in Aristotle, whose writings were two generations earlier than the *Elements*:

> Nor does the geometer suppose falsehoods, as some have said, stating that one should not use a falsehood but that the geometer speaks falsely when he says that the line which is not a foot long is a foot long or that the drawn line which is not straight is straight. But the geometer does not

---

[15] Heath 1908 pg. 200. The assumption of the invariability of figures is usually taken to be implicit in CN 4, and Mueller 1981 points out that Proclus gave a proof of Postulate 4 using only CN 4.

[16] Plato, *Republic*, 510d.

conclude anything from there being this line which he himself has described, but from what is made clear through them.[17]

It is not clear to which geometers Aristotle is referring here, and certainly Euclid never explicitly said anything like this. However, Heath also speculates that part of Euclid's purpose in phrasing his postulates, especially Postulate 1, in the precise way that he did was to emphasize this general point that geometric theorems are not literally true of the physical diagrams.[18] Hence, given that diagrams in Euclidean practice functioned as representations of objects in another domain, how can we use our observations about that practice in I 1 and I 4 to illuminate the relationship they bore to these objects?

We can start by noting that, because the system of the *Elements* definitely does not involve unrestricted diagrammatic inference, it is clear that Euclid must have had some feeling for which properties of the diagram were nonrepresentational, and which ones reliably modeled the corresponding properties of the actual objects about which he was reasoning and were therefore useable in proofs. Let us use this distinction to distinguish two categories of diagrammatically-expressible properties in the domain of the *Elements*. First, there are certain extensional properties of geometric objects which were nowhere explicitly described, but were taken by Euclid to be logically available to the proof, and about which the canonical diagrammatic representation was trustworthy. An example of this would be the fact of the intersecting circles in the equilateral triangle construction of I 1, and the related assertion that the two circles will meet and guarantee the existence of exactly two points, and not one or three or ten. Because Euclid's system includes no axioms which address the existence of the points of intersection between arbitrary geometric objects (except that the fifth postulate refers to the intersection point of two nonparallel straight lines), his justification for the existence of these points in the system of the *Elements* can only come from outside of the system. However, because these sorts of properties are reliably reflected in canonical diagrams which are constructed with reasonable care, and are therefore readily available via simple inspection of the diagram, Euclid routinely passes over them without comment. Beyond the incidence and existence properties exposed by I 1, we can identify several other graphically motivated and diagrammatically reliable prin-

[17] Aristotle, *Posterior Analytics* 76b40. This basic representational stance can also be observed in Leibniz. Manders 1994 pg. 3 quotes Leibniz as saying,"[geometrical] figures must also be regarded as characters, for the circle described on paper is not a true circle and need not be; it is enough that we take it for a circle."

[18] Heath 1908 pg. 195.

ciples whose use was also unquestioned in antiquity. Manders discusses, for example, the positing of the existence of segments and nonempty spatial regions based on diagram features, and implicit assumptions in the *Elements* concerning the relations of inside and outside between diagram parts (*e.g.* that a point is within a circle). Knorr also includes in this category the assumptions implicit in the existence of the "fourth proportional" in the Eudoxan segment calculus of Book 5, as well as a variety of other classical assumptions about the existence of geometric figures of various sizes.[19]

In contrast to these reliable properties, though, Euclidean geometry also recognizes a second category of spatial properties of the domain objects which, although they are also logically available for use in proofs, cannot reliably be "read off" or relied upon based on their diagrammatic representation. Into this category would fall, for example, the inference from a drawn line that it is straight, and more generally inferences involving the precise nature of curves (*e.g.* that a drawn circle is circular), relative sizes of figures and angles (except when one is contained within another), any metrical or proportional data, and the like. Euclidean practice requires that properties like these be attributed by the problem's context and sentential description, rather than by inspection of the graphical properties of the diagram. From a modern point of view, diagrammatic representations of properties of the first category tend to be linked to the ways that the graphical properties of the diagram reflect the overall topology of the represented domain, and importantly, are relatively stable with respect to many kinds of minor perturbations and variations in the drawn figures. (This observation, though in a different context, was one of the things which impressed the nineteenth-century French geometer Poncelet, to whom we will turn in Chapter 5.) On the other hand, representations of second-category properties tend to be more closely associated with the ways in which geometric diagrams or drawings would require graphical exactness in the construction of the objects to which they referred.

For reasoning employing the properties in this first category, it is doubtful that Euclid would have been troubled by explicitly adding the appropriate domain axioms, had someone pointed out their necessity to him. After all, when ancient commentators identified some of these unacknowledged principles in the *Elements*, rather than mount any sort of challenge to them, they simply proposed new postulates and common notions based on them: examples are that a single line will divide a plane, or the above-mentioned principle that two straight lines cannot enclose a planar region.[20] And, while the error in the reasoning was typically identified as a gullible

[19] Wilbur Knorr, personal communication.

[20] See Heath 1908, pp. 232-40.

reliance on the properties of the diagram, under this analysis we can see that this is understandable (and even defensible, given an appropriate diagram semantics) because the relevant domain and diagrammatic properties track each other so well. However, precisely analyzing the problems with reasoning involving diagram based inference with second-category properties would have presented more subtle issues, for which no ancient geometer would have had the necessary theory of symbolism. Complicating this issue still further was the fact that the precise division of properties into those whose expression was reliable and those whose expression was unreliable relative to a set of diagrammatic conventions was not based on any semantic or philosophical theory, but rather on implicit acceptance into the practice of the community of Greek geometers. And, of course, the exact boundaries governing what was to be accepted were somewhat vague, and varied over the years.

Let us consider how proof inferences based on the method of superposition fit into this division. Do they rest on the reliably expressible properties of the first category, or are they associated more with the unreliable ones of the second? First, recall Heath's point that both Euclid's choice of wording of CN 4 and its use in the proof of I 4 entails that coincidence has to be understood as involving the physical idea of moving an object through space and depositing it on top of another. This cannot be interpreted to mean the movement of actual drawn figures, though – the medium of physical diagrams on paper or (as they were in antiquity) on sand does not directly support the movement of diagram parts. In any actual construction involving superposition, figures are not moved *per se*; rather, a copy of the original is produced in the new position, and the original figure is possibly (though not always) deleted from the old, along with the construction marks produced by the movement techniques. Thus, we should understand the diagrammatic practice of duplicating figures used in proofs involving superposition as a way to model this sort of movement in the target domain. This modeling is carried out in the diagram via a particular sort of graphical construction – the duplication of one figure on top of another, along with the erasure of the original – and importantly, the results of superposition are based on noting the coincidence of the relevant parts of the two figures. The construction techniques used are unremarkable; the issue we are concerned with here is how to interpret the diagram resulting from applying the method of superposition, as figure duplication, to an existing diagram. Recall that in *Elements* I 1, the presence of intersecting circles in the diagram reliably signals the existence of the necessary point; in the same way, it appears at first that by physically duplicating one diagram over the marks of another in I 4, the student is invited to conclude not only that the two base lines intersect, but that they *must* intersect at a particular distinguished point. However, unlike the situation with I 1, this conclusion seems to require of the student the ability to perform the various constructions with an arbitrary degree of accuracy, and therefore smacks of illegitimately depending on metrical properties of the diagram. Rather than simply being able to read off a stable, first-category property from a construction, as Euclid routinely does throughout the *Elements*, the reliability of using a diagram to ground results derived from the method of superposition is much more problematic. Specifically, it appears that such an inference would depend in a much more sophisticated way than in I 1 on the fact that certain operations performed on these imperfect representations will reliably track real relations on the underlying domain.

Another way to view the status of superposition is to perform the following thought experiment. Suppose that a careful physical superposition is performed on the actual diagram of I 4, and the relevant features of the diagram (lines BC and EF) could not be made to quite overlay. Would the conclusion be that the diagram was inaccurate, or that the proposed coincidence of the lines is in fact erroneous? Answering the first would place the results of figure duplication in the category of unreliable diagrammatic properties, and answering the second would place the results in the reliable category. Here, I suspect that almost everyone, including Euclid, would be forced to conclude that the diagram was inaccurate, and therefore that exact geometric coincidence cannot be reliably inferred from the properties of any given physical artifact. This, again, is in contrast with the similar situation at I 1. If, contrary to fact, the construction of the circles were to have been done as carefully as possible such that there was no discernable fault with their construction, and yet they still could not be made to intersect, then this fact would presumably have been taken as evidence that the intersection point did not exist at all, or at least as evidence of a fairly significant shortcoming in our understanding of the behavior of circles in the plane.

The preceding discussion suggests a framework for explaining why we might allow the bare intersection of lines in a Euclidean diagram can carry existential force in a proof, while rejecting that coincidence of figures (and hence, congruence) could be also be inferred from figure overlay in the same diagram. This explanation depends on making a distinction between the sorts of domain properties which are expressed by Euclidean diagrams. Only certain of these domain properties are first-category, in the sense that they are accurately modeled by the corresponding properties in the diagram. The other diagrammatically-expressible domain properties are either imperfectly modeled by the conventions of the graphical representation, or cannot be guaranteed to remain accurately expressed through the various legal diagrammatic transformations. However, this framework lacks (among other things) a precise way to determine which properties fall into each category, and so cannot be used to settle all of the issues surrounding Euclid's use of diagrams. Indeed, given the paucity of Euclid's exposition, it seems that a consistently supported account is probably not possible to supply. But, we can see clearly that any account of the legitimacy of inferences involving implicit properties like incidence or coincidence must depend on the details of the view one takes of the relation between geometric diagrams and their referents. And, these details will in turn depend on making a clear distinction between the syntactic representations used in a proof and their semantic grounding – something which has only been possible in the last eighty years or so. Finally, this observation helps to explain another point which we made above. Given that Euclid and his followers probably had only a hazy account of the details of how constructed diagrams were related to their underlying referents, it is not surprising to find that Euclid was reluctant to employ the method of superposition in the *Elements*. Heath speculates that the reasons Euclid employed it were only that it was commonly accepted in the geometric tradition which came before him, and that Euclid could not come up with any other way to prove the relevant propositions.

To review, then, what we can conclude from our discussion of Euclidean diagrammatic practice is that any precise account of the standing of diagrams in any proof procedure must depend on an overall theory of the proof's semantics. As long as the proof is viewed as being about something other than the representational artifacts themselves, in order to justify the various manipulations of the representations which occur in the proof, we need to be able to argue that these manipulations are reflective of actual, truth-preserving relations among the entities to which the representations refer. In the case of superposition, for example, the required figure duplication in the diagram would have been thought to derive its validity from the conception of motion in the diagram's denotata, and the modeling of the resultant figure at the endpoint of that motion. For Euclid and the ancient geometers, the possibility of such superposition movement, and its accurate reflection by figure duplication, must have been ultimately tolerable – after all, Euclid *did* employ the method without comment when it suited him.[21] However, questions surrounding the ultimate legitimacy of allowing movement into geometric proofs troubled many later commentators, and led them

[21] This point is emphasized by Mueller 1981 pg. 23. The ancient geometers, including Euclid, routinely accepted the rigid movement of geometric elements in other proof contexts (see *Elements* I 44), and in the definition of the spirals mentioned above. Whether or not acceptance of geometric techniques involving rigid movement should logically entail acceptance of proof methods like superposition is an interesting question, but not one which I will address.

to consider more deeply how the proof methods of geometry interact with and are dependent on the characteristics of the subject matter. For example, Friedman interprets Kant's philosophy of geometry as resting on the assumption that the judgments of pure spatial intuition, upon which Kant believes subject matter of geometry will ultimately rest, will essentially involve the presence of a set of purely kinematic properties in the transcendental action of the imagination (translation and rotation), and that these properties will ground the very possibility of Euclidean constructions.[22] And, efforts by the nineteenth-century physicist Hermann von Helmholtz to respond to Kant in light of the discovery of non-Euclidean geometry (and, according to Heath, to explain the notion of coincidence in CN 4) led Helmholtz to maintain that these kinematic properties are not given *a priori*, but rather are derived empirically from the subject's own perceptions of bodily movement, and thus that there is a necessary linkage between the truths of geometry and the laws of mechanics.[23] As we will see, the basic intellectual progression shows that as philosophical views on the subject matter of geometry became more abstract over time, the linkage between the Euclidean representing diagrams and the represented domain became more and more tenuous. In order to examine this progression in detail, let us move now to the Renaissance, and see how geometry began to break away from the traditional Euclidean domain of the kind of intuitable, visualizable figures which can be drawn in the sand.

---

[22] See Friedman 1997.

[23] Cf. Heath 1981 pg. 226. In *Metaphysics* 1064a30, Aristotle spoke of mathematics as a science concerned with stationary objects, and in this way contrasted it with physics. Heath also cites Veronese, Schopenhauer, and Russell as philosophers who worried about the conception of motion in geometric construction and proof procedures.

4

Descartes and the Rise of Analytic Geometry

The first major break with Euclidean practice occurred with the seventeenth-century development by Descartes of analytic geometry. Descartes' basic insight was that all the problems of classical geometry would be solvable if one could know the exact length of the construction lines needed, and furthermore that these lengths can be represented without the need for absolute measurements by taking them as ratios of lengths which are relative to a given line segment. At the beginning of Book I of his *Géométrie*, he states this clearly:

> Any problem in geometry can easily be reduced to such terms that a knowledge of the lengths of certain straight lines is sufficient for its construction. Just as arithmetic consists of only four or five operations, namely, addition, subtraction, multiplication, division and the extraction of roots, which may be considered a kind of division, so in geometry, to find required lines it is merely necessary to add or subtract other lines; or else, taking one line which I shall call unity in order to relate it as closely as possible to numbers, and which can in general be chosen arbitrarily, and having given two other lines, to find a fourth line which shall be to one of the given lines as the other is to unity (which is the same as multiplication)...[1]

Descartes next proceeds to give ruler-and-compass methods which result in the ability to construct lines of any length (relative to some given line)

[1] Descartes 1637, pg. 297.

which can be specified by a combination of these five arithmetic functions. Given this, his proof method will be familiar to anyone knowing the meaning of "analysis:"

> If, then, we wish to solve any problem, we first suppose the solution already effected, and give names to all the lines that seem needful for its construction – to those that are unknown as well as to those that are known. Then, making no distinction between known and unknown lines, we must unravel the difficulty in any way that shows most naturally the relations between these lines, until we find it possible to express a single quantity in two ways. This will constitute an equation, since the terms of one of these two expressions are together equal to the terms of the other.2

Commenting on Descartes' approach, Heath has observed that, "the essential difference between the Greek and the modern [Cartesian] method is that the Greeks did not direct their efforts to making the fixed lines of a figure as few as possible, but rather to expressing their equations between areas in as short and simple a form as possible."[3] Descartes' method involved proliferating the construction lines, and hence the derived expressions representing their lengths, in hopes of making explicit as many interrelationships as possible. Basically, his procedure was to arithmetize all of the relations between the lengths of the constructed lines in a problem, and in this way to codify all of these relations into equations. He would next attempt to find those lengths which can be expressed in two (or more) ways, and set the related expressions equal to one another. Then, he would use the methods of algebra to solve the resulting simultaneous equations, his object being to express the lengths of all the necessary lines in terms of the lengths of given lines and the five arithmetic functions. If the equations cannot be solved because there is too little information, then the initial problem does not specify a constructible figure.[4] But, if the algebraic manipulations yielded a solution, then he could use the resultant variable assignments to specify the lengths of each of the lines in the construction, and by using the construction

---

[2] Descartes 1637, pg. 300.

[3] See Descartes 1637, pg. 309, translator's footnote 42. Heath's wording here is unfortunate. He is not referring here to minimizing a set of actual algebraic equations (indeed, there were no such equations in classical Greek geometric practice), but rather to minimizing the number and complexity of equality relationships between the areas of figures.

[4] Descartes puts it this way: "We must find as many such equations as there are supposed to be unknown lines; but if, after considering everything involved, so many cannot be found, it is evident that the question is not entirely determined. In such a case we may choose arbitrarily lines of known length for each unknown line to which there corresponds no equation." Descartes 1637, pg. 300.

techniques with which he began the *Géométrie*, he could give a legal Euclidean construction for it.

By associating each Euclidean construction problem with the solution to a particular set of equations, Descartes gained two closely related advantages over the classical methods. Most obviously, his methods represented a way to radically simplify the intricate and often subtle constructions and diagrammatic techniques demanded by classical practice. Descartes in the *Géométrie* is quite boastful about this aspect of the superiority of his new methods, claiming that,

> If it is remembered that in the method I use all problems which present themselves to geometers reduce to a single type, namely, to the question of finding the values of the roots of an equation, it will be clear that a list can be made of all the ways of finding the roots, and that it will then be easy to prove our method the simplest and most general.[5]

We can contrast this approach with the diagram-driven aspect of Euclidean plane geometry described above. Traditional Euclidean proofs had always involved hybrid diagrams – along with the points, lines, and arcs of the construction proper, during the course of a proof diagrams also often acquired new textual labels, metasymbols such as congruence indicators and parallelism icons, right angle markings, and so on. These sorts of symbols made these intermediate diagrams graphically complex symbols in their own right. Furthermore, as we have observed, all these diagrams always had to be accompanied by the discursive text of the proof, in which important diagrammatic information (ranging from such simple things as segment length equalities, to ratios, parallelism, and the classification of lines as straight or conic) was made explicit. Finally, many Euclidean proofs contain the use by citation of prior results, in which a previously established pattern of construction steps may have to be performed on the current diagram. This in turn involves the (often surprisingly complex) identification and relabeling of diagrammatic counterparts, and importantly, the potential requirement of proving additional cases based on the presence of new diagrammatic elements and construction marks. So, in addition to involving hybrid diagrams in an essential way, full Euclidean proofs were intricately hierarchical in the ways in which the component and subcomponent diagrams related to one another. In contrast, Cartesian methods provided a uniform and simple notation in which to prove geometric theorems.

---

[5] Descartes 1637, pg. 216.

The second and most important advantage that Cartesian geometry held over Euclidean, however, was that Cartesian geometry provided a way to bring the power of general algebraic methods to bear on the structure of traditional proofs. We can see this most clearly by comparing traditional proofs with Cartesian ones. These new proofs were not simple step-by-step translations of their Euclidean counterparts. The algebraic representations used in the new proof practice did not support the same sorts of case distinctions that the diagrams did, and so the proof case structures (and thus, the individuation of proofs) were correspondingly different. Cases which traditional practice had treated as distinct could often be handled all at once when reasoning with equations. A simple example will make this clear. Consider the different cases which arise when considering two infinitely extended straight lines. In the Euclidean diagram-driven practice, this situation automatically gives rise to two cases: the lines will intersect at some point, or that they are parallel. A rigorous Euclidean proof must treat each of these cases separately, although the reasoning and constructions may be similar or identical in each case.[6] However, by using Descartes' equational representations, the required reasoning can often proceed without reference to these cases; the two cases are implicitly accounted for by the existence or non-existence of a simultaneous solution to the equations representing the two lines.[7] Further, the use of algebraic techniques that have no classical geometric analog was often a part of the new proofs. Manders describes one such instance, taken from Book III of Descartes' *Géométrie*:

> Having obtained from a geometrical problem a quartic equation with as coefficients algebraic terms formed from the known quantities of the problem, Descartes first invokes the theory of factorization for the quartic (single literal coefficients). He then re-inserts the compound coefficient terms in the resulting auxiliary cubic, to obtain for this specific problem a ruler-and-compass constructible factorization, which is not available for the

---

[6] Manders 1994, pg. 6, cites a reasonably simple proof by Apollonius as an example of "the highest standard of geometrical reasoning as far as rigor... is concerned." The proof has 87 major cases, some of which spawn their own case structures. Euclid himself was inclined to ignore most of the cases, putting only the most difficult or tricky ones in the *Elements* and leaving the rest as (often implicit) exercises for the reader.

[7] See also Descartes' example in Descartes 1637, pp. 302-4. In this example, Descartes considers the possible intersection points between a circle and a line, and so considers the roots of the equation representing their simultaneous solution. When this equation has two roots, there are two intersection points; when it has one, the line is tangent to the circle; and when there are no real roots, there is no intersection between the figures.

general quartic. Only thus can his method match the ancient result he cites, that the problem is ruler-and-compass constructible.[8]

Propelled by these two advantages – a homogeneous notation and a simplified and more general proof method linked to the traditional Euclidean domain – Cartesian techniques rapidly became the dominant formalism for geometrical research. The popularity of these techniques, however, had effects that went beyond an increase in power of geometric proof methods, and impacted the traditional, intuitable, visually-oriented conception of the subject matter of geometry. Recall that, throughout antiquity and up to the seventeenth century, geometry was conceived of as descriptive of relations between independently existing objects in space, whether Platonic forms or some other idealized version of actual existent entities. The case structure of Euclidean proofs was driven by the possibilities which were dictated by the diagrammatically represented spatial properties of these objects, and so the utility and power of diagrams in Euclidean practice rested on their presumed ability to represent and track the spatial relations into which these objects entered. This tight linkage between the diagrammatic representation and the object domain, however, was made problematic by Descartes' use of equations to represent geometric entities. Algebraic manipulation of equations often involves operations only distantly related to geometric transformations, as Descartes' use of the quartic factorization demonstrates. Further, by mapping the geometric domain of plane figures onto the algebraic domain of equations, analytic geometry showed how it was possible to coherently decouple the representations of geometry from its traditional subject matter. The representations, formal tools, and proof techniques of Cartesian geometry were all drawn from algebra and analysis, making it difficult to maintain that the practitioners were only involved in the study of the properties of plane figures.

In particular, the replacement of Euclidean proof practice – a structured method for modifying some diagram or set of diagrams – with sets of equations solved over the Real plane brought up a set of questions about how this "new" geometry is related to the old. There are three sorts of issues here. Most obviously, by using the new Cartesian notation it became possible to describe and reason about geometric objects (such as complex curves and

---

[8] Manders 1994. We should note that apparently algebraic relations (*e.g.*, CN 1, things equal to the same thing are also equal to one another) were explicitly present in the Common Notions of the *Elements*, and indeed they were necessary for the theory of proportionals developed in the later books. However, in contrast to the algebra used by Descartes, the "things" to which Euclid is referring here are strictly points, lines, curves, and areas, rather than more abstract numerical quantities.

higher-dimensional objects) which were beyond the reach of traditional methods. For example, we find Descartes telling us that those who adopt his approach "will find no great difficulty" in applying it profitably to the study of higher-order curves and spirals, which had been the source of enormous trouble for the ancients.[9] However, this additional mathematical power appeared to some leading nineteenth-century geometers to have been purchased by sacrificing many of the aesthetic advantages of traditional methods, for the reasons outlined above. Chasles, for example, quoted the great analyst Lagrange as conceding that, "there are nevertheless problems in which [synthetic methods] appear more advantageous, partly because of their intrinsic clarity and partly because of the elegance and ease of their solutions."[10] Carnot referred to the "hieroglyphics of analysis" as an impediment to geometric intuition, and Monge and Maclaurin were also strong advocates of synthetic methods in geometry.

This conflict between the mathematical power of the Cartesian methods and the intuitive clarity of the Euclidean ones brought to light a second, more philosophical difference. At the time, geometry held a privileged position in mathematics as the sole subfield whose propositions were thought to be precisely reflective of truths holding in the world. The figures with which geometers worked were seen as straightforward idealizations of physical objects existing in a real, three-dimensional space. Indeed, the absurdity of higher-dimensional spaces had been confirmed by no less than Descartes and Leibniz, and Möbius in 1827 rejected as impossible the concept of superposing two figures in a four-dimensional space because "such a space cannot be thought about."[11] Against this, the functions, derivatives, n-spaces, and other entities studied in algebra and analysis seemed only distantly related to truths about the world, and so results concerning them were less firmly grounded than geometrical ones. These sorts of doubts were not universally held; for example, Gauss had concluded by 1817 that the truth of the Euclidean postulates was rooted in experience, and so "we must place geometry not in the same class with arithmetic, which is purely *a priori*, but with [the empirical science of] mechanics."[12] And, by the end the nineteenth century, Gauss' view had prevailed. Nevertheless, arguments asserting the logical and epistemic priority of geometry over algebra and analysis carried weight at the beginning of the nineteenth century, and contributed to the reemergence of Euclidean diagram-driven geometric practices and the examination of their relationship to the newer Cartesian techniques.

---

[9] Descartes 1637, pg. 317.
[10] See Kline 1972, pg. 835.
[11] Kline 1972, pg. 1028.
[12] Kline 1972, pg. 872.

There was a third issue as well which concerned the interrelationship between these two geometries, related to the expressiveness of the two systems over areas which they had in common. Eighteenth and nineteenth-century developments in geometry gradually showed that Descartes' original premise, that any problem in classical geometry would be solvable by knowing only "the lengths of certain straight lines," was not completely accurate. In particular, the then-new field of projective geometry considered the properties of figures which remained constant when the points of those figures were mapped, or "projected," in various ways onto other figures. The theorems governing these properties were often difficult to express in their full generality with the Cartesian system. A succinct statement of the different sorts of results suited to each geometric system was given by the great French geometer Gergonne in 1826:

> The different theories which constitute the science of extension [geometry] can be placed into two distinct classes. There are, in the first place, certain theorems which depend essentially on metrical relations found to exist between the different parts of extension which one studies, and which, consequently, can be proved only by the aid of the principles of algebra. On the other hand, there are others which are in fact independent of these relations, and follow simply from the positions which the geometric entities upon which we reason have to one another; *and although one may often deduce these latter from theorems of algebra, it is always possible, by proceeding in a suitable manner, to free them from this dependence.*[13]

What this quote shows is that, even though the power of Cartesian methods was well known by this time, there was an identifiable class of geometric results for which the tools of algebra were not well suited. This realization, coupled with the revival of interest in synthetic methods referred to above, led to a sophisticated reexamination of the methods of traditional, synthetic geometry by some of the leading projective geometers during the first part of the nineteenth century. And, although the geometers involved were unsuccessful in restoring the primacy of the manipulation of diagrams in proof practice, their efforts touched on two themes which we shall see again in Part II: first, that diagrams can be freed without loss of rigor from

---

[13] Gergonne, *Annales de Mathématiques*, V. 16 (1826), pg. 209. Translated in Nagel 1939, pg. 148, emphasis added.

their traditional semantics of representing visualizable, intuitable figures; and second, that graphical operations on these diagrams can, in principle, be as general and mechanical as those employed by algebra.

5

Geometric Diagrams in the Nineteenth Century

5.1 Diagrams in the Geometry of Poncelet

These two themes combine in the work of the nineteenth-century French geometer Poncelet. Poncelet was inspired by the early results of projective geometry, and was heavily influenced by the revival of ordinary synthetic geometry started by Carnot. He became interested in attempting to precisely quantify the advantages that algebraic methods held over traditional, synthetic ones in geometry proofs. Poncelet was primarily struck by the previously mentioned fact that Cartesian proof methods did away with many of the often tedious case distinctions in traditional proofs, and attributed this benefit to two closely related characteristics of algebraic methods. First, the operations of algebra were completely general: with the exception of division by zero, any operation could be applied to any quantity-sign. Poncelet contrasted this with the specificity of synthetic proofs:

> While analytic geometry offers by its characteristic method general and uniform means of proceeding to the solution of questions which present themselves ... [and] arrives at results whose generality is without bound, the other [synthetic geometry] proceeds by chance; its way depends completely on the sagacity of those who employ it, and its results are almost always limited to the particular figure one considers.[1]

[1] See Kline 1972, pg. 834.

Although Poncelet believed algebra to be ultimately concerned with individual discernable quantities, mathematicians had by this time successfully introduced signs for negative and imaginary numbers, and generalized the relevant operations so that they were defined over these new signs.[2] For example, in algebra one can always reason with the result of subtracting a quantity *a* from another quantity *b*, without knowing in advance which of *a* or *b* was greater. Traditional diagrammatic practice supported no such general operations; the ability to apply a given construction technique was always dependent on the surrounding context. Second, Poncelet saw clearly that the generality of algebraic operations made it possible that algebraic operations and procedures could be *mechanical*, in the sense that there was no need for an agent to interpret the signs as real quantities in order to correctly apply the reasoning at every step. In traditional proof practice, however, the tightly integrated role of diagrams forced there to be at every step an interpretation of the proof's signs in terms of visualizable figures. In fact, if such an interpretation is not possible, Euclidean reasoning cannot proceed.[3] Poncelet puts this situation succinctly:

> [In geometry,] one always reasons upon the magnitudes themselves which are always real and existing, and one never draws conclusions which do not hold for the objects of sense, whether conceived in imagination or presented to sight.[4]

How would it be possible to include abstract signs and operations in geometric proof practice without translating the problems into the notation of algebra? Poncelet believed, in effect, that the answer lay in treating the entire diagram as a complex but indeterminate sign, rather than as a signifier for some particular, visualizable geometric configuration:

> If it only were possible to employ implicit reasoning, *abstracting from the actual diagram*, and if it were permissible to use the consequences of such

[2] Poncelet refers to these negative and imaginary quantities as "creatures of the brain," and contrasts their corresponding "signs of non-existence" with the presumably more firmly grounded "signs of real quantities." At the time Poncelet was writing, infinitesimal and imaginary quantities were still moderately controversial in mathematics, although they were routinely used. See Poncelet 1865, pp. xi-xii.

[3] An important caveat to this statement concerns the occasional use of diagrams in *reductio ad absurdum* proofs in the *Elements*. See, *e.g.*, *Elements* I 19 (Heath 1908), and several examples in Book III. The strategy here typically involves placing a set of constraints on the diagram which are so strong that no diagram can satisfy them, and reasoning until the impossibility becomes manifest in some way. Poncelet never addresses this point.

[4] Poncelet 1865, pp. xii-xiii.

> reasoning, this state of affairs [the weakness of synthetic methods] would no longer continue. Ordinary geometry, *without using the notation and algorithms of algebra*, would become the rival of analytic geometry, which it indeed already is whenever we have to abandon the framework of explicit reasoning.[5]

Poncelet's task was therefore figuring out a consistent way to abstract from the given diagram. He first observed that Euclidean diagrams were already routinely thought of as indeterminate with regards to exact position of the constituent figures, noting that the diagram's "actual position [is] arbitrary and in a way indeterminate with respect to all the possible positions it could assume without violating the conditions which are supposed to hold between its different parts."[6] This observation that traditional diagrams were always positionally indeterminate led Poncelet to focus on the sorts of continuous graphical transformations which could be applied to these diagrams, while still keeping valid any proofs in which they were used. Again, he was inspired in this by the general applicability of the operations of algebra. For example, in an equation one can always multiply each side by any quantity, and the relationship expressed by the equation will still hold. Poncelet considered the analogous question for diagrams: what operations can always be performed on diagrams, such that any reasoning which employs these diagrams must continue to hold? Starting with the indeterminacy of position in traditional diagrams, Poncelet was led to investigate a general characteristic he believed to hold of all valid positional diagrammatic transformations, which he called the *principle of continuity*:

> The principle of continuity, considered simply from the point of view of geometry, consists in this, that if we suppose a given figure to change its position by having its points undergo a continuous motion without violating the conditions initially assumed to hold between them, the... properties which hold for the first position of the figure still hold in a generalized form for all the derived figures...[7]

---

[5] Poncelet 1865, pp. xii-xiii, emphasis added. The last sentence is a reference to the problems with using algebraic methods in projective geometry about which Gergonne spoke.

[6] Poncelet 1865, pp. xii-xiii.

[7] Poncelet 1865, pg. xiv. Kline points out that this principle is not entirely original with Poncelet; rather, the spirit of it can be identified in the initial development of calculus: "In a broad philosophical sense [the use of principles like this one] goes back to Leibniz, who stated in 1687 that when the differences between two cases can be made smaller than any datum in the given, the differences can be made smaller than any given quantity in the result." Kline 1972, pg. 843.

Basically, the principle of continuity simply restates the method of reasoning *mutatis mutandis* in geometric proofs. It says that, starting from a diagram that satisfies a initial set of conditions, the diagram can be perturbed via a continuous motion of some set of its parts, and if the motion does not cause the diagram to violate the initial conditions, then those same conditions (possibly in a generalized form) will hold in any resultant diagram. Further, Poncelet felt that these sorts of rigid movements over the objects of the diagram would be as mechanical as the operations used in algebra, in that once a particular condition-preserving motion had been specified, it could always be performed without regard to whatever the diagram might be interpreted as referring to. Of course, determining the applicability of this principle would require, for each individual case, a way of precisely describing the initial conditions on the diagram and the types of diagrammatic motion which could be guaranteed not to violate them; without these, the principle would simply have the status of an heuristic. Poncelet supplied some of the details of the way this would be done, although not to modern standards of rigor; the interested reader should refer to the initial sections of Poncelet 1822.

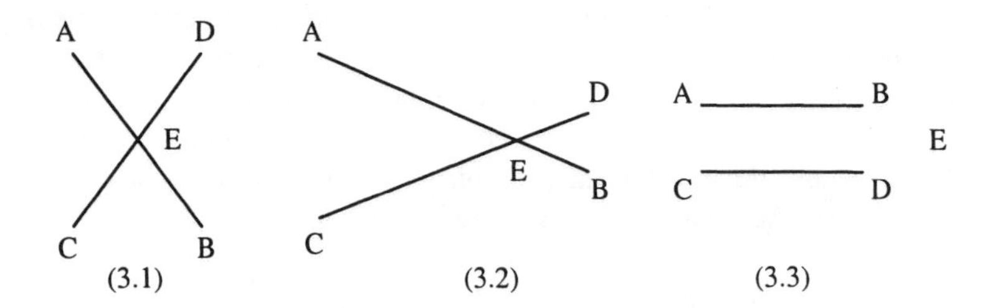

Figure 3: The Principle of Continuity

Thinking about the principle of continuity allowed Poncelet to make a crucial conceptual leap. Consider diagram 3.1 above, representing two infinitely extended distinct lines AB and CD, along with the sole initial condition that AB and CD will intersect at point E. A simple application of Poncelet's principle would allow us to move AB and CD in the diagram so that the angle AEC becomes more and more acute, effectively pushing E further and further from A and C, as shown by diagram 3.2. Poncelet asserted that this transformation (changing the relative angle of AB and CD) respected the principle of continuity because it would not violate any of the conditions initially assumed in the original construction of the diagram, and therefore that any proof which employs diagram 3.1 as a representation for these initial conditions should be equally valid using diagram 3.2, or indeed any of the intermediate diagrams. Poncelet concluded that diagram 3.1 could therefore be coherently taken to be a sign representing a range of possible configurations of lines, varying in the degree of angle AEC, rather than a simple depiction of a single configuration. Further, he realized that the power of this conclusion was not restricted to the role of diagrams in proof methodology. He had taken the first steps toward providing a principled account of how individual *proofs* could be thought of as applicable to different arrangements of objects. By abstracting from specific configurations of objects to the general conditions which give rise to them, he showed how, for instance, a proof could use diagram 3.1 to reason about a range of actual situations, one of which might be ordinarily represented by diagram 3.2.

So far, Poncelet did not depart very much from the traditional interpretation of Euclidean geometric diagrams. But now, suppose that we remove the initial condition that AB and CD must intersect. As we have observed before, classical Euclidean practice would then require that two cases be distinguished in any proof concerning AB and CD: that AB and CD intersect at some point E, or that AB and CD are parallel. Depending on the logical role of AB and CD in the ensuing proof, these two cases might involve identical reasoning, but nevertheless a complete Euclidean proof must include both. This case distinction would be necessary even if our reasoning employs diagram 3.1 as above, to stand for a range of possible intersecting configurations of AB and CD. On the other hand, Cartesian geometric proof techniques would not ordinarily recognize a case distinction here; the equations of the two lines would just have no simultaneous solution if the lines did not intersect, or a single solution if they intersected at a point. Poncelet felt that the relative structural simplicity of a Cartesian proof which lacks this case distinction was a direct result of the ability to use algebraic representations which implicitly included both possibilities. Further, Poncelet saw that through the use of complex numbers – signs which have no interpretation in the traditional, quantity-based interpretation of algebra – the methods of algebra used in analytic geometry allowed one to reason uniformly with *any* two higher-degree equations without having to explicitly consider the number of actual, visualizable intersections that the corresponding curves had.

Poncelet's strategy to collapse these two traditional cases in a synthetic proof relied on the principle of continuity and the analogy to imaginary elements in algebra. He proposed that a graphical representation of the general, unconstrained position of two lines AB and CD would look something like diagram 3.1. As we have already noted, Poncelet's practice allows him to interpret this diagram to also represent situations like that depicted by diagram 3.2. However, Poncelet believed that the principle of continuity also allowed him to continue to move the intersection point E until E is an "imaginary" point "at infinity" (suggested here by diagram 3.3), a situation which would be unintelligible to the ancients, and for which there would be no classical diagram.[8] He held that because of this possibility, diagram 3.1 could also be used in a synthetic-style proof to represent the case that AB and CD were parallel to one another, as long as the reasoning in the proof properly took into account the possibility that the intersection point E might be imaginary. Therefore, if diagrams could be interpreted in this way, no case distinction would be necessary, and diagram 3.1 would be able to represent both the case where AB and CD intersect and the case where they are parallel.

The immediate problem that Poncelet faced was explaining exactly what sort of thing an imaginary point was, and showing how it was related to geometry's traditional ontology of points, lines and planes. For both of these questions, Poncelet invoked the analogy with algebra. Just as negative and imaginary numbers have no interpretation in terms of quantities, so this imaginary intersection point of two parallel lines would have no interpretation in terms of the traditional objects of geometry. However, just as such numbers are used in algebraic reasoning and are implicitly defined by the rules of deduction into which they enter, so Poncelet's imaginary points should be able to be used in geometric reasoning. Poncelet therefore set about giving a system of principles for geometric reasoning involving such imaginary points. Enthused by his success at this, he also went on to employ the principle of continuity as justification for introducing various other imaginary elements into his geometry, along with occasional diagrammatic conventions to represent them.[9] In this way, he was able to secure for

---

[8] The astronomer Kepler, as part of his 1615 investigation of the properties of conic sections (note this is pre-Descartes), was apparently the first to introduce the notion of points "at infinity" where parallel lines would meet. Kepler's use of these points was not thoroughly worked out, however, and his idea was apparently ignored by professional geometers. See Anglin 1994, pg. 158, and Fishback 1962, pg. 27. I have found no evidence that Poncelet was aware of Kepler's work. We should also remember that Poncelet's conception of the infinite was still quite primitive by modern mathematical standards.

[9] A less straightforward example of Poncelet's imaginary elements was the positing of two "general" points of intersection between two circles, even when the circles were not drawn to actually intersect in the usual way. These points can then be used to define a common "chord" for the two circles, which has many of the geometric properties of its traditional cousin. Because in Poncelet's system this line can be defined whether or not the circles actually intersect, certain types of case distinctions in traditional Euclidean proofs can be eliminated. See Manders 1994 for a fuller explanation and comparison to Euclid's proofs in *Elements* III 35-7. Also see Kline 1972, pp. 843-4.

proofs in his synthetic system much of the structural simplicity which characterized the analogous Cartesian proofs.

Let us step back from the details of Poncelet's theory of imaginary elements to consider some larger implications. What Poncelet did with his introduction of imaginary points, although he did not fully recognize it, was to suggest a radical change in the interpretation of geometric diagrams, and thereby an equally radical change in the accepted subject matter of geometry. He proposed, in effect, that diagrams be treated as complex signs whose denotata remain constant when operated on by transformations licensed by the principle of continuity. The core of this idea – that the relation between drawn diagrams and their denotata is not one-to-one – is familiar from our discussion of Euclid in chapter 3, but Poncelet developed it in a much more sophisticated way. And, importantly, he proposed that these denotata need *not* be a traditional, visualizable configuration of figures; they were more analogous to abstract groups of related conditions. This was the major break from the Euclidean and Cartesian geometric traditions. Recall our observation above that Euclidean techniques were based, from proof step to proof step, on the ability to construct diagrams which directly depict the figures described (or, in the case of a *reductio* proof, demonstrate graphically that a figure is impossible), and as Poncelet observed, "one never draws conclusions which do not hold for the objects of sense." Cartesian techniques were similarly tied to an ontology of visualizable, constructible objects: the purpose of the algebra was to yield equations composed of simple algebraic functions, which could then be used to construct lines of certain lengths. And, although the intermediate algebraic manipulations could make use of functions and quantities which did not have ruler-and-compass constructible correlates (such as higher-degree polynomials with imaginary roots), a proof's preconditions and conclusions, and therefore the range of its possible subject matter, always had to be ruler-and-compass constructible. However, Poncelet's diagrams are best interpreted simply as graphical notations for general sets of constraints over a system of objects, which may or may not be coherently interpretable as classical geometric figures. These constraints were the conditions which were given to hold at the beginning of the proof, and consisted of things like the existence of figures, explicit intersections, and interior and exterior facts. And, importantly, the geometric objects which were thereby constrained were not required to be intuitable or visualizable.

Viewed in this way, Poncelet was not simply proposing a new proof technique for traditional geometry. By introducing an imaginary point at which parallel lines intersect, he is not giving a new property of parallel lines as traditionally understood; in fact, Euclid's Book I definition of parallel lines specifically states that they do not meet each other. Rather, he is best understood as redefining the semantics of geometry proofs, and constructing a new geometric system which was patterned after the traditional one and employed the traditional vocabulary in suggestive ways. As I have said, though, Poncelet himself was not fully aware of the radical nature of what he was proposing, and he rapidly became embroiled in a long debate with Cauchy and Hankel over the philosophical status of objects whose existence was predicated on the operations of the principle of continuity. From a philosophical viewpoint, of the three methods for introducing new mathematical objects – simple postulation, implicit definition via the operations into which the new objects enter, and construction out of already-accepted objects and functions – it was unclear where these new objects fell. Nagel's observations confirm that the use of implicit definition was an important part of Poncelet's strategy to philosophically ground his imaginary elements:

> Just as the "imaginary numbers" in algebra are not an additional kind of numerical magnitude homogeneous with the familiar cardinals, but are simply signs "implicitly defined" by the rules of operation into which they enter, so the "imaginary points" generated by the principle of continuity are also uninterpreted signs for something or other, whose range of denotation is limited only by the rules of combination specified for them.[10]

In places, however, Poncelet also seems to claim that these imaginary entities have an independent existence on a par with classically-understood geometric entities. Perhaps because he was less interested in philosophical matters, Poncelet never clearly set out his views on these sorts of ontological questions, and the legitimacy of Poncelet's precise formulation of the principle of continuity and the entities it licensed remained controversial among other nineteenth-century geometers.[11] However, the immediate effect of Poncelet's work was to stimulate a debate on the status of geometric systems and the objects to which they refer, similar to the debates which were sim-

---

[10] Nagel 1939, pg. 159.

[11] Subsequent developments in projective geometry ended up vindicating Poncelet's general approach, if not the details of his development. Contemporary projective geometry includes "imaginary" non-Euclidean entities of the sort envisioned by Poncelet, although the interpretation of the terms denoting these entities is typically done within the framework of a formalist philosophy of mathematics which would have been foreign to him. In brief, the real projective plane is defined by augmenting traditionally inspired points and lines with new points (called ideal points) and one new line (the ideal line). Under these expanded definitions of point and line, it can be shown that any two points will lie on a single line, and any two lines will meet in a single point. See Fishback 1962 for details.

mering over the status of complex numbers in algebra and the use of complex coordinates in projective geometry. These debates paved the way for Hilbert's completely axiomatic development of geometry, to which we will turn in chapter 5.

Poncelet's investigation of the principle of continuity had another important effect on the geometry of the day. Roughly contemporaneously with Poncelet, Gergonne had noted that many major projective theorems had *duals*: associated theorems in which the terms "points" and "lines" (for plane geometry) or "points" and "planes" (for solid geometry) were interchanged. Gergonne illustrated this by constructing two axiom systems, duals of each other, in order to give two equivalent developments of a piece of projective geometry called the theory of reciprocal polars. In one, he took the standard Euclidean/Cartesian position that points were the fundamental objects, and defined lines and planes based on points. In the other, he took planes as fundamental, and defined points and lines in terms of various intersections between planes. For each theorem of one system, there was a corresponding dual theorem in the other, and their proofs were related in a systematic way.[12] Poncelet, in the course of thinking about the status of his imaginary objects, had also noticed this dualism in the theory of reciprocal polars, and began a long debate with Gergonne over priority in its discovery. And, while it now appears that Gergonne was the more perceptive of the two, the heat generated by this controversy focused attention on the development of dual systems in geometry, resulting in the work of Plücker.

Plücker's contribution to geometry did not involve diagrams *per se*; he was mainly interested in dualities which arise between different equational systems in analytic geometry. Specifically, he first showed that, in place of the traditional development of plane analytic geometry in terms of points in a two-dimensional space, it was possible to construct an equivalent but dual geometry where conic sections were the fundamental objects, and points were defined by the intersections of the conics. In this case, the space (or manifold) in which the geometry is done is considered to be five-dimensional, because five coefficients are required to specify the general equation of the conic. Plücker went on to generalize this work to show rigorously that dualities will exist between any pair of objects having the same dimensionality relative to the chosen manifold. In a powerful sense, the geometries generated by these dualities will express the same subject matter – every theorem will have corresponding duals – and yet the representations in the different geometries (including any pictorial or diagrammatic representation) will necessarily appear very different. Also, unlike Poncelet,

[12] See Nagel 1939, pp. 181-183.

Plücker recognized the philosophical import of his work. Instead of the older, Platonic view of geometry as concerned with a set of simples against which diagrams and proof statements have a fixed interpretation, Plücker realized that geometry, like algebra, could be conceived as a pure science of structure, divorced from any particular interpretation, and concerned with investigating a certain set of abstract patterns which can manifest themselves in many different representations. Further, he held that the validity of these patterns is not a consequence of their traditional geometric character: "every geometrical relation is to be viewed as the pictorial representation of an analytic relation, which, irrespective of every interpretation, has its independent validity."[13]

So, through the work of Poncelet and Plücker, the traditional subject matter of geometry, and consequently the role of diagrams in geometric proof practices, came under heavy attack. Poncelet's work, while attempting to increase the power of diagrammatic methods, showed that it was possible to coherently posit new geometric objects which did not have interpretations in the traditional scheme, and thereby helped geometry free itself from investigating only objects which could be visualized and drawn. Plücker went further and showed rigorously that, at least in analytic geometry, the dimensionality of the representation with which the geometry was carried out was mathematically arbitrary. After Plücker and Poncelet's work, it was clear that the objects of geometry need not be interpreted in terms of visualizable configurations of primitive points and lines, as they had been since Euclid. Therefore, one of the traditional roles for diagrams in a geometry proof – to provide a fixed interpretation of terms over an accepted and independently-understood domain – had become quite problematic. This conventional interpretation, however, still carried a sort of epistemic primacy among geometers. It took two final events to erase this primacy and cement the decline of diagrams: the invention of non-Euclidean geometry and the associated reexamination of foundational issues; and the subsequent development of axiomatic techniques and the rise of pure (uninterpreted) geometry. Section 5.2 will take up the first of these, and as such will address certain larger issues in the philosophy of mathematics relevant to the history of geometry. In section 5.3 we will return to the main stream of geometric developments, and show how the philosophical changes chronicled in section 5.2 meshed with the technical achievements of Pasch and Hilbert to give rise to a new attitude toward the subject matter of geometry and the role of diagrams in proof.

---

[13] Plücker, *System der Geometrie des Raumes*, 1846, pg. 322. Translated in Nagel 1939.

5.2 Non-Euclidean Geometries and the Rejection of Kantianism

Against the background of the work of Poncelet and Plücker, the mid-1800s invention of non-Euclidean geometries, in which Euclid's fifth postulate was modified in various ways, seems at first to be an almost inescapable consequence. Poncelet had exhibited a coherent and useful geometrical system in which certain of the basic objects and the relations into which they entered were not visualizable. Plücker had built on this and shown that the linkage between these basic objects and their traditional representation was only conventional, in the sense that the same abstract geometrical relationships could be conceptualized (via the choice of specific theorems from the possible duals) using multiple distinct ontologies and their correspondingly distinct representations. The effect of these developments, as we have observed, was to challenge the classical conception of the subject matter of geometry, and to lay some of the necessary groundwork for studying geometric structures on their own, independent of any required interpretation. The creation of non-Euclidean systems of geometry, with their implicit rejection of the traditional visualizable subject matter of Euclid, appears at first to be a natural and obvious extension of this trend, and hence a technical achievement which mathematicians would have little trouble accepting.

However, history does not support this explanation. The development of non-Euclidean geometries instead grew out of a separate legacy of research, on Euclid's parallel postulate, and was not especially linked to the nineteenth-century developments in projective geometry just discussed. We will briefly summarize this research now. Historically, although the geometrical principle it expressed was invariably accepted to be true, Euclid's original formulation of the parallel postulate in the *Elements* was thought by many geometers to have a significant shortcoming: it was too conceptually complicated, and as a consequence the principle it expressed was suspected to be less fundamental than his other four postulates. Heath translates Euclid's postulate in this way:

> That, if a straight line falling on two straight lines makes the interior angles on the same side less than two right angles, the two straight lines, if produced indefinitely, meet on that side on which are the angles less than the two right angles.

Two general strategies were employed to address this perceived shortcoming in Euclid's formulation. The first consisted of efforts to show that the parallel postulate could be replaced in the system of the *Elements* by something theoretically simpler and more logically primitive. By 1800, several logically equivalent axioms had been discovered, including the Playfair formulation that is now used in many elementary geometry texts. Because success in this strategy was measured directly by the perceived simplicity of the replacement axioms, it is clear that this strategy had been at least moderately successful by the beginning of the nineteenth century.[14]

The second, more ambitious, and more popular approach was to attempt to show that the parallel postulate itself was logically eliminable from the system of the *Elements*, because it was a consequence of Euclid's other axioms and common notions. This sort of proof was attempted very early in the history of the *Elements*, at least as early as Ptolemy. In his commentary on the *Elements*, Proclus exposed the fallacy in Ptolemy's proof, and attempts one of his own. He motivates his own proof of the postulate by observing that because there exist lines which can converge but never meet (as in conic sections), the axiom that converging straight lines will meet is something which is only plausible, and therefore it does not have the quality of necessary self-evident truth which would qualify it for placement among the initial postulates of the system:

> So in this case, when the two right angles are lessened, the fact that the straight lines converge is true and necessary, but the statement that they will meet sometime, since they converge more and more as they are produced, is [only] plausible, but it is not necessary in the absence of some argument showing that this is true. It is a known fact that some lines [*i.e.*, curves] exist which approach each other indefinitely, but yet remain non-intersecting; this seems improbable and paradoxical, but nevertheless it is true and fully ascertained with regard to other species of lines. .... [U]ntil the statement in the postulate is clinched by proof, the facts shown in the case of the other lines may direct our imagination the opposite way.[15]

However, the proofs that were given by Proclus and others were always later revealed to depend on some further geometric principle or principles which were not contained in the *Elements*. Proclus's proof, for example, can be

[14] Playfair's axiom states that given a point and a line not containing the point, there is one and only one line containing the point which is parallel to the given line. Several other equivalent axioms, including versions by Wallis, Laplace, Lambert, Legendre, Bolyai, and Gauss, are given in Eves and Newsom 1965, pg. 60.

[15] Quoted in Eves and Newsom 1965, pg. 59.

shown to depend on a tacit assumption that parallel lines are always a bounded distance apart.

The first glimmerings of non-Euclidean geometry can be found in the exhaustive effort to prove the parallel postulate by Saccheri in early 1700s. Saccheri was the first geometer to attempt a full-scale attack on the postulate by using the *reductio* method of proof. And, although the contradiction which Saccheri finally claimed to have found was spurious, his use of *reductio* methods inspired Lambert and Legendre in their efforts to prove the postulate, and the body of theorems which they derived based on the denial of the parallel postulate was considered more or less independently by Gauss, Loebachevsky, and Bolyai in the early 1800s. The first papers seriously proposing the distinct mathematical study of non-Euclidean geometry (for reasons both obvious and, for our purposes, provocative, Loebachevsky called it "imaginary geometry") were published independently by Loebachevsky and Bolyai around 1830. Gauss, however, was the first to recognize the broader philosophical implications of Loebachevsky's and Bolyai's non-Euclidean geometry, suggesting in correspondence as early as 1817 that these new non-Euclidean systems were not only logically consistent, but could be potentially be the actual geometry of physical space, and therefore constituted a challenge to the accepted theoretical primacy of Euclidean geometry.[16] By way of emphasizing this point, Gauss at one point even set about measuring the angle sum of a triangle formed by the peaks of three local mountains, in order to empirically determine the truth of Euclid's parallel postulate. (His experiments were inconclusive; the potential measurement error made the results consistent with either of his hypotheses.) Ironically, though, because the frontier of geometric research was centered around many of the issues in projective geometry discussed in the previous section, work on non-Euclidean geometry was largely ignored until the great German mathematician Riemann took it up in a series of lectures and papers in 1854.

Even after Riemann's prestige focused attention on the possibility of non-Euclidean geometry, however, its acceptance as something more than a logical curiosity took almost forty years, involved the passing of a generation of mathematicians, and required some fundamental changes in the philosophical underpinnings of geometry and of axiomatic mathematics in general. The principal reason why this process of acceptance was so lengthy

---

[16] Gauss writes in a letter of 1817: "I am becoming more and more convinced that the [physical] necessity of our [Euclidean] geometry cannot be proved, at least not by human reason nor for human reason. Perhaps in another life we will be able to obtain insight into the nature of space, which is now unattainable. Until then we must place geometry not in the same class with arithmetic, which is purely *a priori*, but with mechanics." Gauss, *Werke*, 8, 177. Translated in Kline 1972, pg. 872.

and difficult is related to the clear intuitive conviction, unquestioned by the work of any mathematician of the previous 20 centuries, of the literal truth of the Euclidean axioms relative to the objects populating the natural world. This conviction was so strong that it spawned several attempts during the seventeenth and eighteenth centuries to ground the (presumably more problematic) basic theorems of algebra and analysis in the principles of the *Elements*, and thereby secure for these subjects the same degree of certainty as was accorded to the perceived "facts" of geometry.[17] Even during the early nineteenth century, geometers as a group still firmly believed that they were at root involved in the precise, scientific study of the actual extensional properties of objects, and that their abstractions of indivisible points, breadthless lines, and perfectly flat planes were simply the necessary theoretical tools for this study. Further, Euclidean geometric intuition (*i.e.*, the ability to visualize the interactions between geometric objects in the plane or in ordinary space, consistent with the propositions of the *Elements*) was thought to be both a legitimate and a necessary guide to these extensional properties.

Evidence for the currency of this essentialist view on geometric intuition in the early nineteenth century is not difficult to find. Influential philosophers such as Hobbes, Locke, and Leibniz all explicitly held that the truth of the Euclidean axioms was guaranteed by the basic structure of the world. Even Frege, in the course of his 1884 investigations into the foundations of elementary arithmetic, held that, "the truths of geometry govern all that is spatially intuitable, whether actual or the product of our fancy," and that, "in geometry, it is quite intelligible that general propositions should be derived from intuition."[18] Poncelet's work on geometry was inspired by the possibility of combining the cognitive advantages of the Euclidean synthetic method, with its reliance on intuition and visualization as methods of discovery and proof, with the technical advantages of algebraic techniques. A particularly frank example of this view on the primacy of Euclidean intuition can be found in the correspondence between Möbius and Aplet about the work of Grassman, an influential geometer who came after Poncelet. Aplet complains that "an abstract theory of extension [*i.e.*, geometry] such as Grassman wishes, can be developed only from concepts; but the source of mathematical knowledge is found not in concepts, but in intuition." Möbius agrees, observing that Grassman's book "keeps itself too much aloof from all intuition, which is the essential trait of mathematical knowledge."[19]

---

[17] Kline 1972, pg. 862.

[18] Frege 1884, §13-4.

[19] See Nagel 1939, pp. 173-4.

The most compelling evidence for the general acceptance of the reliability of Euclidean intuition, however, can be found in the reluctance of geometers to consider results holding for spaces of dimension higher than three. This is a classical requirement found in Aristotle's *De Caelo*, and also present in the work of Cardan, Descartes, Pascal, and Leibniz.[20] A nice illustration of the force of this restriction on early nineteenth century geometrical thought can be found in an 1827 discussion of superposition, again by Möbius. Klein writes that:

> [Möbius] pointed out that geometrical figures that could not be superposed in three dimensions because they are mirror images of each other could be superposed in four dimensions. But then he says, "Since, however, such a space cannot be thought about, the superposition is impossible."[21]

Even after the study of non-Euclidean geometry was made fashionable by Riemann in 1864, orthodox Euclideanism about physical space was still the rule. Frege took up the question of four dimensional geometry in 1884, suggesting that such investigations occupied a somewhat uneasy middle ground between Euclidean intuition and what he termed "conceptual thought":

> The wildest visions of delirium ... remain, so long as they remain intuitable, still subject to the axioms of geometry. Conceptual thought alone can after a fashion shake off this yoke, when it assumes, say, a space of four dimensions or positive curvature. To study such conceptions is not useless by any means; but it is to leave the ground of intuition entirely behind. If we do make use of intuition even here, as an aid, it is still the same old intuition of Euclidean space, the only one whose structures we can intuit. Only then the intuition is not taken at its face value, but as symbolic of something else; for example, we call straight or plane what we actually intuit as curved. For the purposes of conceptual thought we can always assume the contrary of some one or other of the geometrical axioms, without involving ourselves in any self-contradictions when we pro-

---

[20] Aristotle, *De Caelo* 268a9: "A magnitude if divisible one way is a line, if two ways a surface, and if three a body. Beyond these there is no other magnitude, because the three dimensions are all that there are, and that which is divisible in three directions is divisible in all." See also Kline 1972, pg. 1028.

[21] Klein 1972, pg. 1028.

> ceed to our deductions, despite the conflict between our assumptions and our intuition.[22]

Finally, in 1883, we find the great British mathematician Arthur Cayley claiming in a major lecture that Playfair's formulation of the parallel postulate, "does not need demonstration but is part of our notion of space, of the physical space of our own experience ... which is the representation lying at the foundation of all external experience," and that such space is "if not exactly, at least to the highest degree of approximation, Euclidean space."[23]

The conviction of the metaphysical truth of the Euclidean axioms was reinforced by the dominant philosophy of mathematics of the nineteenth century, Kantianism, which entailed that it was simply *a priori* that the objects of experience would conform to Euclidean intuitions and metrics. According to Kant, the totality of our spatial experience is mediated through a set of categories, through which our faculty of apperception is able to give unity to the original manifold of sense-data. These categories structure the underlying transcendental space in which all of our thinking and perception occurs – what Kant calls our *metaphysical* space – and therefore define the basic parameters under which our pure geometric intuition can operate. Kant held that the organization of the metaphysical space which results from the application of the categories places limits on the activity of imagination in representing geometric forms to our consciousness. Further, our ability to come to know certain features of this space is critical. For example, it is the transcendental knowledge of the infinity of metaphysical space, given to us by reflection on our inner sense, which ultimately justifies the ability of a specific geometrical space to accommodate iterated constructions, and which therefore supports the production of figures in *a priori* geometric imagination of any constructible size or ratio of sizes. Further, Kant holds that the very possibility of Euclidean constructions in geometric imagination is a consequence of our primitive transcendental ability to create a unity out of iterated motion in metaphysical space.[24] In particular, the ability of our imagination to synthesize a line segment out of a succession of points, and to synthesize a circle from a rotation of a line segment about a point, are for Kant proofs *simpliciter* for the logical possibility of segments and circles. And, because Kant held that the axioms and theorems of Euclidean geome-

[22] Frege 1884, §14.

[23] Cayley 1898, v. 11, pp. 429-59.

[24] According to Kant, this ability to represent motion to ourselves is at the heart of the concept of succession, and therefore also is critical to our experience of time. See, *e.g.*, 2nd. ed. Deduction, B154 (Kant 1787). The role of the transcendental synthesis of the imagination in Kant's theory of geometry is discussed in detail in Friedman 1997.

try are simply the elaboration of the laws governing this primitive productive ability in *a priori* geometric imagination, the metaphysical status of Euclidean geometry is founded directly on the conditions of *a priori* imagination.

Kant's approach to general metaphysics therefore secured in one stroke both the necessary character of Euclidean geometrical truth and the consequent applicability of Euclidean geometry to the world of our experience. Further, it entailed that geometry was a qualitatively different sort of science than, say, biology, because there could be no conceivable experiment which could challenge the basic truth of the Euclidean axioms.[25] In this way, Kantianism represented a significant change from the older Platonic realist view, because it located the truth of the Euclidean axioms in conditions on the necessary prerequisites for sensible experience (*i.e.*, intuition), rather than in the presumed ability to use our facilities of reason to divine the properties of a group of external, independently-existing forms. However, by the end of the nineteenth century, the popularity of Kant's philosophy of mathematics had succumbed to pressures from three areas: the direct effects of developments in non-Euclidean geometry, advances in other parts of mathematics which cast doubt on the reliability of geometric intuition, and a general trend toward reductionism and rigor in analysis. We will use the remainder of this section to expand on these themes.

The invention and early development of non-Euclidean geometry showed that there were at least four non-equivalent alternative possibilities for the axioms of geometry – Euclidean (or parabolic) geometry, single- and double-elliptic geometry, and hyperbolic geometry. The relative consistency proofs that soon followed demonstrated that any contradiction in one of the three non-Euclidean geometries would also be derivable within Euclidean geometry. (At the time these proofs were given, it was presumed as a matter of course that Euclidean geometry was itself consistent, for Euclidean geometry was believed to be the actual geometry of the world, and there could be no contradiction inherent in the world's basic fabric. Frege refers to this idea when he says that, "from the truth of the axioms it follows that they do not contradict one another."[26]) By 1880, Klein had shown that these four metric geometries can be thought of as different spe-

---

[25] Kant also gave *a priori* status to the principles of Newtonian mechanics and, as will be observed, to the principles of Aristotelian logic. Both of these theories have been repudiated in some way by subsequent advancements in their science, thus giving wicked substance to Peirce's 1877 comment that, "one may be sure that whatever scientific investigation shall have put out of doubt will presently receive *a priori* demonstration on the part of the metaphysicians." See Haack 1974, pg. 27.

[26] This quote appears in Frege's correspondence with Hilbert. See Frege 1980, pg. 37.

cializations of (non-metrical) projective geometry, and had laid out the modern transformation viewpoint detailing the relationships which all of these geometries bear to one another. Essentially, this viewpoint characterizes an individual geometry as a group of algebraic transformations over entities defined by sets of real-valued coordinates, and studies the structures which are invariant under those transformations. Klein demonstrated that these four geometries differ from one another only in their notion of congruence (*i.e.*, in the functions used to measure distances and angles), and Helmholtz and Poincaré showed that the intuitive notion of a rigid motion could be preserved in three of them. These various technical developments challenged Kant's claim of the metaphysical priority of the Euclidean axioms, and importantly, recalled Gauss's question of the appropriate geometry of the physical world. The very possibility of this question entailed that any claim of the intrinsic truth of Euclidean geometry in the world was more a matter of physical science and the structure of current physical theories than of *a priori* mathematical intuition.

This awareness of the possibility of non-Euclidean geometries led many mathematicians of the late 1800s to revisit the technical foundations of Euclidean geometry, and attempt to reconstruct it in a way that would be more in keeping with the standards of rigor of the time. This research was not the result of any real debate aroused by Gauss's question; mathematicians on the whole remained quite skeptical of the physical reality of non-Euclidean geometries. Riemann speculated that the actual geometry of the world must be at least "locally Euclidean" in his 1854 paper, and Klein and Poincaré viewed non-Euclidean geometries primarily as logical constructions.[27] The provision (by Riemann, Klein, and Beltrami) of Euclidean surface models upon which the various two-dimensional non-Euclidean geometries could be interpreted certainly helped the acceptance of non-Euclidean geometries, but it was not until the use of non-Euclidean geometry in the theory of relativity in the early 1900s that this tradition of established Euclideanism about the nature of physical space was finally diminished. Instead, work on the foundations of Euclidean geometry was spurred because the development and classification of non-Euclidean geometries had made clear that mathematicians had ignored logical shortcomings in Euclid's system. We will discuss aspects of this work related to diagrams in section 5.3. For our present purposes, though, it is important to see that this gradual realization that Euclidean geometry did not occupy a privileged theoretical position also made many mathematicians dubious of the reliability of the *a priori* Euclidean geometric intuitions central to Kant's phi-

[27] Kline 1972, pp. 921-3 and 1033.

losophy of mathematics, and in an ironic reversal of the trend of previous centuries, led to the influential calls by Klein, Peano, Hilbert, and others to found geometry entirely on arithmetic and analysis.[28]

In addition to the invention of non-Euclidean geometry, there were several other developments in other branches of mathematics which also caused trouble for the accepted Kantian view about the reliability of Euclidean geometric intuition. Besides Plücker's work on algebraic transformations of the manifold, we will only mention two others, both concerned with problems in the intuitive notion of a curve. In 1837, Wantzel demonstrated the impossibility of the general angle trisection in Euclidean geometry by showing that the boundaries of Euclidean-constructible figures in the real plane were not actually continuous, but rather contained an infinite number of "holes."[29] In the 1860s, Weierstrass gave several famous examples of functions which were continuous and yet nowhere-differentiable, showing that curves could be defined which exceeded the boundaries of accepted geometric intuition. At first, the sorts of "pathological" functions which defined these curves were difficult for many mathematicians to accept, probably because of the challenges they posed to the usefulness of such intuition. Poincaré himself complained about many of these new functions, protesting that:

> Logic sometimes makes monsters. For half a century we have seen a mass of bizarre functions which appear to be forced to resemble as little as possible honest functions which serve some purpose.[30]

Nevertheless, by the 1880s, it was clear that reliance on geometric intuition as a trustworthy source of knowledge in mathematics had resulted in problems in analysis as well as in geometry.

Finally, the need for a new, more abstract philosophy of mathematics also fit well with the general emphasis in later nineteenth century mathematics on revisiting foundational questions. In algebra, the efforts of British mathematicians such as Peacock, Hamilton, and Cayley had succeeded in

---

[28] See Kline 1972, pg. 1013. See also Friedman 1997 for a discussion of the anti-Kantianism in the philosophy of geometry developed by Helmholtz, Poincaré, and Weyl. The attempt to ground the metaphysical certainty of mathematics in the truth of arithmetic and analysis was fairly quickly realized to be problematic; already by 1887 Helmholtz had suggested that the basic arithmetic concepts of definite quantity and equality were only approximately related to the world of experience.

[29] Wantzel defined the Euclidean subfield of the reals by closing the rationals over the operation of extracting real square roots, and showed that the general angle trisection required points which were outside this Euclidean subfield.

[30] See Kline 1972, pg. 973.

refocusing research in algebra away from the manipulation of magnitudes and onto the more formal and abstract notion of an algebraic structure; and in analysis, Riemann, Cauchy, Weierstrass, and their followers were busy with their projects of rigorizing the calculus and the subsequent arithmetization of analysis. The main intellectual thread running through this research was a brand of mathematical reductionism, whose object was to redefine the subject matter of mathematics as the increasingly rigorous elaboration of a smaller and smaller set of basic notions, and the focus was on defining the elements of this set for various branches of mathematics. These elements were typically members of the real number system and the functions which could be defined upon them. Importantly, though, the choice of these elements as ontological simples carried real philosophical weight. Improvements in understanding of the real numbers and their functions allowed mathematicians to conceptualize them as increasingly complex and elaborate structures definable in ordinary arithmetic. And, it was hoped that arithmetic would be able to supplant geometry as the branch of mathematics whose theorems were most directly reflective of truths in the world, and therefore could be used as a ground for the truth of mathematical statements. Further, analytic methods in geometry suggested a way in which this certainty could be imported back into geometry, even in the face of non-Euclidean theories. In this overall project of reductionism to arithmetic, then, Kant's view of the truths of geometry as given *a priori* to the mind and embedded in the conditions of experience itself was completely unsatisfactory.

One of the first major methodological results of all of these developments was an explicit call, by Frege, Bolzano, Cantor, Dedekind, and others, for the removal from the notion of mathematical proof of any methods or entities whose justification was based on the Kantian-inspired intuitions of (Euclidean) space and time.[31] This was a comprehensive program, encompassing algebra and analysis as well as geometry. Dedekind, for example, writes:

> In saying that arithmetic (algebra, analysis) is only a part of logic I wish to state that I hold that the number concept is completely independent of the ideas or intuitions of space and time, and that I hold it to be an immediate consequence of the laws of pure thought.[32]

---

[31] See Frege 1884, §§40-41. See also Hallett 1994, pp. 158-60.

[32] Dedekind, *Essays on the Theory of Numbers*, pg. 31, translated in Hallett 1994, pg. 158.

However, the precise delineation of the *methods* to be used in this research program (*i.e.*, the definition of the boundaries of the accepted proof methods in mathematics) received its initial development with the work of Frege. Dedekind's reference in the quote above to the "laws of pure thought" as the basis for the concept of number is suggestive, because it implies a (logicist) belief that the methods of logic would be a more satisfactory and reliable grounding for the concepts of mathematics than the Kantian intuitions which he is rejecting. Frege's remarks in his 1884 *Foundations of Arithmetic*, however, make clear that this project is committed to the careful removal of intuition from the procedures of arithmetical *proof*, as well as from the concepts upon which these procedures operate:

> A single [proof] step is often really a whole compendium, equivalent to several simple inferences, and into it there can still creep along with these some element from intuition. In proofs as we know them, progress is by jumps, ... [and] the bigger the jump, the more diverse are the combinations it can represent of simple inferences with axioms derived from intuition. .... [My] demand is not to be denied: every jump must be barred from our deductions.[33]

Frege's comments demonstrate that he is interested in going beyond clarifying the foundations of the concept of number. Unlike, for example, Cauchy's rigorization of the notion of limit in calculus, Frege was interested not only in regrounding a familiar concept, but also in analyzing the methods by which he can validly prove propositions based on this concept. His position was that any principles used in arithmetic proof must be grounded in the self-evident "laws of thought" which would be applicable to any reasoned discourse. For Frege, the theorems of arithmetic are *a priori*, and so must be the case that their proofs, "can be derived exclusively from general laws, which themselves neither need nor admit of proof."[34]

Of course, Frege was not a geometer, and was not particularly interested in geometry except as a source for occasional examples. His work was consistent with the emerging anti-Kantianism in the philosophy of mathematics, and was primarily concerned with investigating the objects and proof procedures used in elementary arithmetic. With Frege 1884 and Frege 1893, he hoped to demonstrate that arithmetic could be developed as a consequence of a set of preexisting logical laws and definitions, and hence that the truth and applicability of its statements would be unassailable. Ultimately, then,

---

[33] Frege 1884, §§90-91. See also §§75-84.
[34] Frege 1884, §3.

arithmetic could be used as a suitable foundation for most other branches of mathematics. Frege, however, did *not* believe that theorems of geometry could be grounded in his account of arithmetic; he cautions that, "we shall do well not to overestimate the extent to which arithmetic is akin to geometry," and that Kant's classification of geometric theorems as synthetic yet *a priori* "revealed their true nature."[35] Frege's arguments for this position were not his most sophisticated; he appeared to simply reject out of hand the notion that the concepts of geometry could be defined without appeal to external (*i.e.*, essentially psychological) intuitions. Further, his reasons for allowing mathematical proofs to make use of *a priori* laws of thought, and not equally *a priori* geometric intuitions, appear to be theory-driven at best. At least, without a careful analysis of such geometric intuitions of the sort he does not provide, it is not clear how strongly he can support his exclusion of intuition-based proof methods in geometry (*e.g.*, proofs which make use of diagrams or other sorts of spatial devices) from the domain of acceptable proof techniques in mathematics.

Nevertheless, for our purposes Frege's work illustrates an important intermediate point in the way that conceptions about the grounds for certainty in geometrical statements evolved away from Kantianism and toward more modern views. Recall that initially the truth of geometric statements was thought to be tied to their application to external entities, typically either Platonic forms or via some type of empiricism. Kantian philosophy of mathematics relocated the grounds for the (still self-evident) truth of Euclidean geometry into the mind, into concepts which are given to us *a priori* as part of the preconditions to experience, and in this way claimed to address both the problem of the Cartesian demon and Hume's skepticism about the possibility of necessary truth. With Frege, we have seen a third position emerge, which is that foundations of arithmetic can be viewed as the consequences of the working out of a set of completely general and content-free laws of thought, and therefore to the extent that the truths of mathematics are based on arithmetic, these truths are analytic rather than synthetic. Frege's explicit exclusion of geometry from his arguments is somewhat unfortunate, especially in view of the developments in analytic geometry chronicled previously. In the next section, we will see how the work of Pasch and Hilbert demonstrated that geometry could be built out of a set of self-contained definitions interpretable within the real number system, and

[35] Frege 1884, §14 and §89. Frege believed that the truths of arithmetic and any mathematics derivable solely from them would turn out to be analytic, in contrast to the presumably synthetic propositions of geometry. This view about the ultimate source of geometric intuitions did not, however, prevent Frege from thinking that a satisfactory geometric proof system could be given. See our discussion of Frege in chapter 10.

so finally how the theorems of geometry themselves could be incorporated into the rest of mathematics and therefore viewed as analytic. And, as we will argue, this shift in philosophical attitudes toward geometry had profound consequences for the role of diagrams in geometric proof.

5.3 Pasch, Hilbert, and the Rise of Pure Geometry

Let us return more directly to developments in geometry. As we have just observed, the general loss of confidence in the use of intuition in proofs in analysis made it important to supply geometry with a more stable foundation than Kant's mysterious and problematic categories. We have just seen how the work of Plücker and the invention of non-Euclidean geometry made the idea that the Euclidean axioms were reflective of absolute truths of physical space at least theoretically problematic. Furthermore, the discovery of non-Euclidean geometry and the recognition of the benefits of rigorization of algebra and analysis had fostered an attempt to rework Euclidean geometry into a more technically satisfactory system, incorporating current mathematical theories and addressing all the various deficiencies which had been noted over the centuries. The work of Klein referenced above had set the stage for this effort, by exposing a powerful and elegant underlying connection between the various metrical geometries and projective geometry, and showing how the then-new theory of algebraic invariants allowed results in the different geometries to be interpreted in terms of developments in other parts of mathematics. The culmination of this reworking was the rigorous and purely axiomatic treatment given in Hilbert's 1899 *Foundations of Geometry*, which we shall discuss below.

For the purposes of our argument, the major effect of this reworking was that the late nineteenth century witnessed the final decline in research on diagrammatic and other synthetic methods in geometry of the sort that Poncelet had envisioned. Of course, the interpretation of diagrams in proofs in non-Euclidean geometries had always been awkward at best, typically involving the implicit use of one of the Euclidean surface models in which the straight lines of the geometry had to be represented as curved in the diagram. However, the characteristic use of diagrams in ordinary Euclidean geometry, with their cognitive basis in the representation of experiential objects in space, was also open to several sorts of criticisms. Because the workings of these diagrams in this practice were dependent on human intuitions about the possibility of certain operations in physical space, and because these intuitions were assumed to be grounded in the specific nature of our experience and psychology, the integral use of diagrams in geometric proofs was viewed as a prime source of fallacies, inexact reasoning, and appeals to unarticulated assumptions. Moreover, the reliability of diagram operations was seen as linked to our ability to trust our spatial intuitions, and therefore allowing diagrams into mathematical proofs could be seen as tantamount to illegitimately importing psychology into mathematics. Consequently, removing psychology from geometry necessitated the elimination of diagram-based moves from the proof procedure, and the rebuilding of geometry on the basis of "logical" deduction alone.

The German geometer Moritz Pasch was the first to clearly articulate this as a research program. For him, eliminating intuition from geometric proof involved structuring the proof system so as to avoid any dependence on the content of the terms involved. In this way, he was following in the intellectual footsteps of Poncelet, who supplied a complete set of rules in order to regulate the use of imaginary geometric elements in a proof. In his influential 1882 book, Pasch gave a particularly clear statement of his view with regards to the use of diagrams in geometric proofs:

> In fact, if geometry is to be genuinely deductive, then the process of inferring must always be independent of the sense of geometrical concepts, *just as it must be independent of diagrams*. .... In the course of a deduction, it is certainly legitimate and useful, though in no way necessary, to think of the reference of the concepts concerned. Indeed, if it is necessary to do so, then the inadequacy of the deduction is revealed, and even the insufficiency of the proof method...[36]

Pasch's work stopped short of a completely formal, symbolic treatment of geometry. He explicitly acknowledged that the primitive terms and propositions of his system were inspired by ordinary experience (Pasch referred to these as "nuclear"), and furthermore held that the empirical truth of all of his theorems would be founded on the truth of the nuclear propositions, which were in turn confirmed by the observation of actual interacting bodies. In this way, he saw himself as employing the same foundational strategy as Euclid, whose actual definitions of "point," "line," and "plane" are not used in the proofs of theorems in the *Elements*, but serve mainly to suggest the correct intuitions and linkages to physical objects. For Pasch, for example, "a physical body is to be called a 'point' if its subdivision into

[36] Pasch 1882, *Vorlesungen über Neuere Geometrie*, emphasis added, translated in Hallett 1994, pg. 160.

parts is incompatible with the limits set by actual observation."[37] However, Pasch almost immediately layered new meanings onto his nuclear terms: *e.g.*, taking an "extended point" to be the set of all rays emerging from a common center, and employed these extended meanings in order to prove most of the significant theorems of his system. Pasch felt that this process of abstraction from a set of empirically-grounded nuclear propositions was typical of mathematics, writing that, "the attempt to bring the maximum number of configurations within a common point of view, so that they may all be considered in a uniform way, has progressively forced the 'proper' meanings of the elementary geometrical terms into the background."[38] He also formally demonstrated the principle of duality in projective geometry first noted by Poncelet and Gergonne, and argued that this principle shows that projective proofs must be *de facto* independent of the denotations of their terms.

By blending empiricism and formalism in this way, Pasch was the first to show in detail how geometry could coherently be split into "pure" and "applied" subfields, and to explain how individual geometric theorems could be useful and express regularities about the world, and yet have proofs which are independent of any kind of experience. For Pasch, pure geometry emerged as a strictly demonstrative discipline which is independent not only of intuitive methods and techniques in proofs, but indeed of *any* effect of the reference of the terms involved, such as unarticulated properties or restrictions. In pure geometry, there is no distinguished subject matter or model in which the terms can acquire independent meanings; the terms are simply variables whose interactions are governed by rules which themselves make no reference to a subject matter. At most, the terms of a geometry are implicitly defined by the axioms which govern them. Thus, the validity of a deduction is strictly a matter of checking for adherence to these previously specified rules. Pure geometry becomes applied when its undefined terms are given an interpretation, as with Pasch's nuclear terms and propositions, and also, although Pasch did not fully appreciate it, when the pure deductive methods that are used can be shown to be truth-preserving in this interpretation. Because, for Pasch, any use of diagrams was intrinsically rooted in spatial intuition, no valid deduction in pure geometry could involve diagram-based operations, and because applied geometry was simply a *post hoc* interpretation of terms arising in pure geometry, it was therefore the case that no deduction in applied geometry could legitimately involve dia-

[37] Nagel 1939, pg. 194. Note the similarity between Pasch's definition and Euclid's, who defines a point to be "that which has no part."

[38] Nagel 1939, pg. 196.

gram-based operations either, although such a deduction could always be interpreted to be about geometric relations in specific diagrams.

The level of formalization and rigor of Pasch's work was very influential, and set the stage for what is probably the most influential contemporary book on pure geometry: Hilbert's 1899 *Foundations of Geometry*. Although Hilbert ironically chose to begin its introduction with an epigraph from Kant's first *Critique* emphasizing the epistemic priority of intuition – "All human knowledge thus begins with intuitions, proceeds thence to concepts, and ends with ideas" – his book has been employed as the basis for a formalist philosophy of mathematics which is almost completely opposed to Kant's. This book, initially aided by the prestige of its author, was immediately hailed as a landmark, and went through seven German editions before Hilbert died. Hilbert was quite familiar with Pasch's work, and improved on it in many ways. Probably the most significant of these improvements over Pasch, in terms of ensuring the book's impact and accessibility, were that Hilbert constructed his axiom system so that it would be recognizably based on the traditional geometry of Euclid, instead of adopting Pasch's strategy of using "proper" points and lines, and that he eschewed the obscure symbolism invented by Pasch in favor of a more straightforward proof notation. However, the main intellectual importance of Hilbert's book is that it demonstrates how pure geometry can be developed without even the thin veneer of empiricism put on it by Pasch. Although in the introduction Hilbert claims only to be concerned with giving a "logical analysis of our perception of space," and he even organizes the presentation of his axioms around intuitive spatial concepts, it is clear very quickly that nothing whatever hangs on Hilbert's choice of the terms "point," "line," and "plane." The opening words of Hilbert's book illustrate the contrast with Pasch's attempt to ground his nuclear terms in the experience of physical bodies:

> Consider three distinct sets of objects. Let the objects of the first set be called *points* and be denoted by A, B, C, ...; let the objects of the second set be called *lines* and be denoted by a,b,c, ...; let the objects of the third set be called planes and be denoted by α,β,γ .... The points, lines, and planes are considered to have certain mutual relations and these relations are denoted by words like *lie, between, congruent.* . The precise and mathematically complete description of these relations follows from the *axioms of geometry.*[39]

---

39 Hilbert 1899, §1.

While it is clear that Hilbert, like Pasch, held that the study of geometry was originally inspired by the our observations of regularities in the measurement of bodies in world, the system he develops in the *Foundations* is designed to be completely self-sufficient and free of any connection to the property of extension.[40] This is an important point, as it represents a complete reversal from the classical view that geometry was the branch of mathematics which was closest to the world, and was only possible because of Pasch's previous separation of pure geometry from applied. There is a great deal of evidence that Hilbert had this sort of independence specifically in mind when he was working on the *Foundations*. In 1884, for example, he wrote the following about his evolving geometric system:

> In general we must state: Our theory furnishes only the *schema* of concepts connected to each other through the unalterable laws of logic. It is left to human reason how it wants to apply this schema to appearance, how it wants to fill it with material.[41]

Later on, in 1899, in a letter Hilbert wrote to Frege concerning the possibility of defining geometric terms via axioms, he said:

> Instead of 'axioms' you can say 'characteristic marks' if you like. But if one is looking for another definition of, *e.g.*, 'points', perhaps through paraphrase in terms of extensionless..., then I reject such attempts as fruitless, illogical and futile. One is looking for something where there is nothing.[42]

Hilbert expanded on and confirmed this view in a set of lectures on geometry in 1921, in which he speaks generally about the independence of axiomatic theories from possible interpretations:

> Through this mapping [of a domain-specific concepts on to terms], the investigation becomes completely detached from concrete reality. The theory has nothing more to do with real objects or with the intuitive content

---

[40] There is one interesting exception to this: the geometry of Hilbert 1899 is limited to three dimensions. In §2, Hilbert explicitly acknowledges this, writing that, "Axiom I 7 expresses the fact that space has no more than three dimensions, whereas Axiom I 8 expresses the fact that space has no less than three dimensions."

[41] Translated in Hallett 1994, pg. 166, emphasis added.

[42] Frege 1980, pg. 41. (The ellipsis is in the original.)

> of knowledge. It is a pure thought construction, of which one can no longer say that it is true or false. Nevertheless, this framework has a meaning for the knowledge of reality, in the sense that it presents a possible form of actual connections. The task of mathematics is then to develop this framework of concepts in a logical way, regardless of whether one was led to it by experience or by systematic speculation.[43]

These quotes also show Hilbert's basic philosophy of mathematics to be significantly different from Frege's. While Frege's work, with which Hilbert was familiar via correspondence by 1895, is predicated on a particular fixed set-theoretic interpretation for the notion of number, and it is critical for Frege's philosophical project that this interpretation be a consequence of the laws which govern the general behavior of concepts, Hilbert's geometrical system was constructed to be extremely flexible in the sorts of interpretations which it allows. Hilbert's basic system does not even assume that the underlying objects must be representable as sets of real numbers; while his axiom of line completeness guarantees that there will be an isomorphism between the points on a line and the real numbers, and thus that the geometry defined by his other axioms will be equivalent to Cartesian geometry over real-valued coordinates, none of his proofs of the theorems of Euclidean plane geometry depend on the presence of this axiom.[44] (Hilbert uses the axiom of line completeness and the Archimedean axiom (his two axioms of continuity) in order to justify constructing models of the geometrical terms which interpret them as functions and real numbers, so that he can prove metatheorems which give independence results for the various groups

---

[43] Translated in Hallett 1994, pg. 168.

[44] Remember that Euclidean plane geometry is not equivalent to two-dimensional real-valued coordinate geometry; Descartes' restriction of his final construction equations to those containing only the four elementary arithmetic operations plus the extraction of square roots had fallen away by Hilbert's time. Hilbert placed two axioms in his continuity group: the Archimedean axiom and the line completeness axiom. In §9 of Hilbert 1899, Hilbert shows that the Cartesian geometry defined by Descartes' five construction operations is an interpretation of his first four axiom groups plus the Archimedean axiom. The second axiom of the continuity group (the axiom of line completeness) is not used in this proof. He also shows in §17 that the Archimedean axiom is unnecessary if the target geometry is restricted to points and straight lines in a plane, and thus that its interpretation is restricted to linear functions. §§36 and 37 discuss the sorts of geometric constructions which can be performed with and without assuming that a compass is available: without the Archimedean axiom, Hilbert's geometry is limited to ruler-and-scale constructions, and cannot in general employ a compass. All the theorems of Euclidean geometry (*i.e.*, theorems about the ruler-and-compass constructible domain) can be established without assuming the postulate of line completeness. But, for example, the axiom of line completeness would be necessary in order to trisect an arbitrary angle in Hilbert's geometry.

of axioms. Hilbert showed that the required manipulation of magnitudes and proportions in his geometry could be handled via an internal calculus of segments based on Pascal's Theorem, rather than requiring a detour through algebra. This feature of Hilbert's development is parallel to Euclid's use of the Eudoxan theory of proportionalities.) In eliminating all external requirements on the subject matter of his geometry, Hilbert went considerably beyond Frege's immediate goal of the removing psychological influences from mathematical proofs, and placed himself much more in line with Pasch's ideas about axiom systems than with Frege's. Unlike Pasch, though, we have seen from the quote from the *Foundations* above that Hilbert did not even attempt "nuclear" definitions for his six primitive terms. These six terms are only implicitly defined by his axioms, and even though he mentions in *Foundations* §1 that his axiom groupings are inspired by "certain related facts basic to our intuition," nothing at all hangs on this declaration. In fact, in 1905 Hilbert wrote in one of his notebooks that, "the points, lines, and planes of my geometry are nothing other than things of thought, and as such have nothing whatsoever to do with real points, lines, and planes."[45]

In the specific geometric system he gives in the *Foundations*, Hilbert implicitly adopts Pasch's attitude toward the use of diagrams: that their use necessarily entails selecting from all the possible dual systems one particular interpretation for the primitive terms of the geometry (in the historical case, this interpretation is structured around traditional points and lines instead of, *e.g.*, intersecting conics or sets of rays), and that any proof which incorporates such an interpretation in its methods is therefore suspect. Hilbert was aware of many of the logical controversies surrounding the system of the *Elements*, including the difficulties with the method of superposition and Euclid's lack of axioms regulating the incidence and existence properties among its figures. We have earlier argued that Euclid neglected these issues in part because of the power of his canonical diagrammatic representation. In Hilbert's system, only proofs which proceed strictly from the (sentential) statements of the axioms and hold themselves aloof from all interpretations of their primitive terms can be known to be legitimate and free from possible unarticulated assumptions imported by a suggestive interpretation. Thus, *e.g.*, the method of superposition is untenable in this system, because it requires that geometric figures be interpreted to stand for bodies in space and obey certain laws of rigid motion. Hilbert therefore replaced Euclid's treatment of congruence with a set of five postulates which specify the essential properties of the congruence relation completely axiomatically. This

[45] Translated in Hallett 1994, pg 167.

stance against any sort of traditional diagrammatic, figure-based interpretation can be seen explicitly in a lecture on geometry Hilbert gave in 1894, in which he equates the use of diagrams in a proof with "experimental geometry:"

> A system of points, lines, planes is called a diagram or figure. The proof [of the theorem he is discussing] can indeed be given by calling on a suitable figure, but this appeal is not at all necessary. [It merely] makes the interpretation easier, and it [the appeal to diagrams] is a fruitful means of discovering new propositions. Nevertheless, care, since it [the use of figures] can easily be misleading. A theorem is only proved when the proof is *completely independent of the diagram.* The proof must call step by step on the preceding axioms. The making of figures is [equivalent to] the experimentation of the physicist, and experimental geometry is already over with the [laying down of the] axioms.[46]

From this quote, it is obvious what Hilbert thought about the relative importance of experimental and axiomatic geometry.

In addition to sharing Pasch's view about the characteristics of pure geometry, where interpretations of terms (such as those provided by diagrams) are forbidden, Hilbert's work went further than Pasch and provided a more powerful reason for eliminating diagrams from geometric proofs. Hilbert had a more fully-developed position than Pasch on the proper role of logic and proof in his system, possibly spurred by his correspondence with Frege in the late 1890s. We will explore Hilbert's philosophy of logic more extensively in Part II, where we will concentrate specifically on the evolution of diagram-based systems for logic. Here it will be sufficient to give a brief

---

[46] Translated in Hallett 1994, pg. 162, emphasis added. We can contrast Hilbert's position on experimental geometry with the views of Roger Bacon:

> [T]here are two modes of acquiring knowledge, namely, by reasoning and experience. Reasoning draws a conclusion and makes us grant the conclusion, but does not make the conclusion certain, nor does it remove doubt so that the mind may rest on the intuition f truth, unless the mind discovers it by the path of experience.... This is also evident in mathematics, where proof is most convincing. But the mind of one who has the most convincing proof in regard to the equilateral triangle [Bacon is here referring to *Elements* I 1] will never cleave to the conclusion without experience, nor will he heed it, but will disregard it until experience is offered him by the intersection of two circles, from either intersection of which two lines may be drawn to the extremities of the given line; but then the man accepts the conclusion without any question.

Bacon 1928, pg. 583. Clearly, Bacon would disagree with Hilbert that geometry is made obsolete by the laying down of the axioms; for Bacon, the "making of figures" is precisely what imparts the quality of certainty to the proofs of geometry.

outline of the position Hilbert had arrived at by the 1920s. Basically, by emptying geometric terms of any reference to a privileged subject matter, Hilbert had thereby guaranteed that the only possible grounds for moving from one statement in a proof to another would be the operation of logical laws over the structure of the representations used in the geometry. Diagrams could not be an appropriate symbol system for these representations, because the proofs which relied on diagrams could not meet Hilbert's minimum metalogical criteria for any deductive system for mathematics. The symbols used in geometric proofs had to be sentential in order that the logic which justified the proofs in which they appeared would be able to qualify as proper in Hilbert's view. More precisely, Hilbert required that the representation associated with each proof step had to consist of a finite number of distinguishable and identifiable signs, and the logical principles which licensed the deduction of one step from another had to consist only of basic arithmetic ("calculating") operations on these signs:

> The role of the language in the expression of the logical connections between thoughts corresponds to the sign language in calculation. In following a logical passage of thought with the help of this logical language, we carry out simultaneously a calculation, in which manifold logically elementary processes are put together according to practiced rules.[47]

Hilbert's recognition of a linkage between logic and calculation is clearly not novel; the substance of this quote could have been lifted almost directly from Leibniz's musings about the *ars combinatoria*. What Hilbert contributes is a much more sophisticated understanding of the capabilities of logic, plus the recognition that this program could be, if not as globally applicable as Leibniz had hoped, at least useful at the foundations of mathematics in regulating the scope of mathematical proof. At any rate, the effect of Hilbert's new geometric practice, with a symbolism patterned on language and an explicit central role for formal axiomatics and logic, was to finally cement the eclipse of the diagram as a logically effective tool in geometric proofs.

While Hilbert's views on the requirements which logic places on the symbolism of proofs seem today to be fairly routine and straightforward, it is important to keep in mind that Hilbert and Frege were among the first thinkers to lay out the modern conception of the role of logic in mathematical proof, and Hilbert was the first to see the utility of the mathematical study of logical calculi themselves (*i.e.*, metamathematics). Stemming from

---

[47] Translated in Hallett 1994, pg. 180.

the technical advances of Boole, Frege, and Russell, logic in the early 1900s had evolved to the point where it was quite plausible to believe that all of pure mathematics could be derived from the repeated application of the rules of logic over a set of axioms. Further, within Hilbert's formalist program, the simple consistency of an axiomatic system was sufficient to philosophically ground the mathematics which was developed from it.[48] The way in which proofs in pure geometry were to be integrated into this overall framework seemed now to be clear: stemming from the unifying algebraic models of Klein and the axiomatic development of Pasch and Hilbert, geometric proofs would be treated as essentially algebraic demonstrations concerning the possible abstract structures which could be defined by a particular set of axioms. The consistency of Hilbert's geometry would be derived from its linkage to real number theory via his two continuity postulates. Interestingly, though, there is a place where Hilbert appears to have been of two minds on whether sufficiently precise geometric diagrams could function as an operand for a group of appropriately-tailored logical laws. In the midst of an early (1900) discussion of the role of symbols in proof, Hilbert even suggests that "...arithmetical signs are written diagrams, and geometrical diagrams are drawn formulas," implying that at an essential level, these symbols have the same representational qualities, and thus that the same sorts of operations would be appropriate to each.[49] However, Hilbert later retreated from this possibility. Hallett, relying on an unpublished 1921 lecture, claims that Hilbert:

> suggests that geometrical figures are different in so far as they are taken to have properties that cannot be described in purely discrete (and thus not in finitary) terms. They are therefore not amenable to the primitive intuitions of the *finite Einstellung*, unlike elementary arithmetical and proof "figures."[50]

In this way, Hilbert completely barred the door on the formal use of diagrams not only in geometry, but in all of mathematical proof.

Given Hilbert's prestige and the success of his *Foundations*, most professional geometers quickly became convinced of the power of viewing their

---

[48] Gödel's 1931 incompleteness theorem showed that it was not possible to use the simple logic Hilbert permitted in these sorts of consistency proofs to show the consistency of full number theory, and thus that Hilbert's program would not work for anything but a very restricted subset of arithmetic.

[49] See Hilbert 1900.

[50] Hallett 1994, note 45. The lectures referred to were given by Hilbert in the winter semester of 1921 at the Mathematisches Institut, Georg-August Universität, Göttingen.

subject as an uninterpreted symbolic system regulated solely by a set of sentential axioms, and research turned to developing various models of the system and clarifying their relations to other branches of mathematics. Prominent early examples of this research included Minkowski's use of a slight modification of Hilbert's axiom system in work in number theory, and the development of the various geometries which arise by selecting coordinates out of a restricted algebraic field. The presence of this trend in research allows us to understand Mueller's assertion with which we started this part of the book: that by the twentieth century, "traditional or descriptive geometry is simply an interpretation of certain parts of modern algebra." Correspondingly, reliance on figures and construction techniques became largely absent in formal geometric proofs. In 1915, Pasch characterized the prevalent state of affairs in geometric proof in this way:

> A statement B1 is a consequence of B only because its derivation from the latter is completely independent of the meanings of the geometrical concepts occurring in it, so that the proof can be carried through without support from an actual or imagined diagram or from any sort of "intuition." A proof which does not meet this condition is no mathematical proof.[51]

This definition of acceptable proof in geometry is still current today.

---

[51] Pasch, *Mathematik und Logik*, 1915, pg. 37, translated in Nagel 1939, pg. 196.

6

Summary

What lessons can we draw from our investigation of history of geometry? Recall the task which we set for ourselves at the beginning of this part of the book: to explain the decline and eventual eclipse of diagrammatic methods in formal geometric proofs, given that diagrams were so important in early geometric practice and have undisputed cognitive advantages. In the course of addressing this question, we have traced the development of geometric methods from ancient procedures in which diagrams and constructions played a critical logical role, through the introduction of Cartesian techniques and the first linkages with algebra, and into the modern axiomatic era in which formalisms borrowed from algebra and analysis are the primary representations. More interestingly, though, we have seen in this history how the accepted subject matter and philosophical framework of geometry itself underwent a parallel evolution. It is the conclusion of Part I of this book that these two types of change – in geometric proof methods and in the interpretation of those methods – combined to account for the decline in importance of diagram-aware deductive systems in geometry today.

Let us review the main points of this evolution of the perceived subject matter of geometry. The basic philosophical progression can be understood in terms of the attempt to account for the apparent necessary applicability of geometric theorems to the objects of the physical world. Geometry as a discipline originated in the need to solve problems concerned with distances and areas in surveying and cartography. Its subject matter was therefore the physical features of the world, and the logical relationship its conclusions bore to these features was therefore contingent, akin to that of any physical theory. The foundational contribution of the Greeks was to restructure geometry into a quasi-independent and axiomatically-based science of the abstract extensional properties of idealized objects in space. Crucially, though, these extensional properties modeled by the geometry were just those that actual physical objects were assumed to possess: they had definite boundaries, they could be equal in shape to other objects, they could either touch or not touch, they could be decomposed into parts, etc. In this way, the theorems of geometry were linked to the behavior of objects in the world. Euclid's reliance on diagrams, constructions, and superposition as important tools in the practice of this geometry was consistent with this view. And, as geometry developed away from its early empirical roots, the relation between the objects of experience and the objects of the geometry - came to be conceived in terms of a set of Platonic forms, whose properties would underwrite the truth of geometric theorems, and whose necessary linkage to the objects of the physical world explain the usefulness of these theorems. This basic account persisted through Descartes' introduction of algebraic techniques into the proof methods of classical geometry. These techniques did not (at first) affect this basic Platonic connection; Descartes merely provided a way to employ the newly-invented tools of algebra in geometric investigations. Geometry in the seventeenth and eighteenth centuries was still uniformly conceived by its practitioners as the science of extension, and its theorems and diagrams described the extensional relationships which all objects would necessarily bear toward one another.

After Hume's critique of the problems surrounding knowledge of necessary truths about physical objects, though, this basic Platonic conception needed to become more sophisticated. Kant's philosophy of geometry improved on Plato's by basing the Euclidean axioms on the pure intuitions of space and time necessary for consciousness itself. In this way, Kant explained how we can come to have *a priori* knowledge of geometry, and reestablished the theorems of geometry as both universal and necessary. However, Kant's achievement came at a time when the techniques of geometry were changing rapidly, and the two centuries surrounding his work saw a split gradually arise between the traditional visualizable approach to the subject matter and the range of methods which were developed to explore that subject matter. The new Cartesian techniques provided a way in which the study of geometry could directly benefit from tools and theories being developed in the rest of mathematics, and gave rise to a powerful geometric proof practice in which the primary representations did not always have to correspond to well-behaved, visualizable groups of intuitable entities. Further, too-literal reliance on the properties of diagrams had been recognized since antiquity as a significant source for error, and Descartes' algebraic model for traditional geometry offered an elegant solution to this problem. This methodological trend away from an intuitable subject matter was accelerated by Poncelet's work, which although aimed at reinvigorating traditional synthetic geometric methods, went beyond Descartes' representations and proposed a coherent and useful geometry in which certain of the geometric entities themselves were not visualizable. These developments had the effect of moving the actual practice of geometry away from its Greek roots as the abstract science of the extensional properties of the objects of experience, and in this way made increasingly problematic the use of diagrams and constructions in this practice.

This tension between geometric methods and ontology was resolved by the recognition of consistent non-Euclidean systems of geometry and the subsequent abandonment of the Kantian view of geometry. After this discovery, it was unclear whether the theorems of geometry could even be considered to be *true* of objects in the world, let alone descriptive of their necessary properties, because of the uncertainty about the world's actual geometry. This uneasy philosophical situation was remedied by the work of Pasch and Hilbert, which formally distinguished pure geometry from its applied subfield, and thereby detached mathematical research in geometry from any concerns about the reality of objects which it describes. Pure geometry was conceived of as inspired by problems in the measurement and modeling of physical entities, but (by fiat) completely independent from them. Applied geometry, on the other hand, would still face these philosophical questions, and in this way would be no different from any other scientific theory, but would not be an area of mathematical research.[1] Pursuant to this goal, Pasch and Hilbert rebuilt pure geometry on an axiomatic foundation whose properties were formally independent of any interpretation of its terms, and in particular, were not exclusively linked to spatial intuitions. Moreover, Klein's work on the classification of different geometries showed how geometric structures were related to algebraic structures, and the continuity postulate of Hilbert's geometry formalized the linkage between real number theory and the constructs in pure geometry. As a result of these developments, then, pure geometry was explicitly constructed so that research

[1] This device of explicitly dividing mathematics into pure and applied subfields has always seemed to me suspiciously like a failure of philosophical nerve on the part of the mathematicians of the late nineteenth century. Instead of facing the foundational questions squarely, they simply defined them away, and created a division of labor in which any questions of philosophy were somebody else's problem. The success of this split had the effect of removing these foundational questions from the set of issues considered distinctly mathematical. We can see that the intellectual effects of this split continue today: because issues at the foundations of mathematics are not generally thought of as primarily mathematical, they are rarely viewed as issues about which the opinions of research mathematicians should be specially privileged, or indeed issues with which research mathematicians should be concerned at all.

in it could always be carried out without the use of diagrams, via the symbolism and methods of algebra.

Setting aside the convenience and power of algebraic methods, though, there are two interrelated reasons why diagrammatic methods could not be used in proofs in pure geometry. The first and simplest reason is a result of the requirement of indeterminacy of interpretation that is central to pure geometry. Traditional geometry diagrams, with their conventional graphical representations of points, lines, and planes, force a particular interpretation on the geometric vocabulary, and so diagram use in proofs thereby invites dependencies on properties of the particular interpretation which might not be shared by all of the possible dual interpretations. In order to foreclose this possibility, the justification for each step of a proof in pure geometry must trace back solely to the axioms. And, because of the nature of the underlying logic, those axioms and the deductive relations into which they enter can make no mention of diagrams or other similar entities. Therefore, in no way could a diagram or its properties affect the logical relationships occurring in such a proof.

The second reason why diagrams are not found in proofs in pure geometry is more subtle, and is a result of the consensus among nineteenth-century mathematicians that proofs in any sort of pure mathematics be free of any dependency on facts unique to our particular psychology. Pure mathematics, although concerned with structures created solely by the mind, was to be established on the basis of methods which only appealed to the most basic and fundamental symbolic capabilities of the mind. By the early 1900s, Hilbert and his followers had taken this restriction to mean that proofs had to be finite in both length and alphabet, and involve only deductive operations which were based on syntactic, pattern-matching operations over discreet symbols of that alphabet. In particular, no semantic interpretation over the symbols of the proof would be allowed to affect the deductive validity of any step of the proof. Against this standard, traditional diagram use in geometry is problematic: the kinds of inference which are classically performed based on diagrams (*e.g.*, constructions, superposition arguments, and conclusions involving incidence and existence) seemed to these mathematicians to be essentially tied to our particular cognitive makeup, perceptive ability, and spatial intuition.

So, by the 1920s we can see that proofs in pure geometry must reject diagram use for reasons connected both with the representational function of diagrams, and with the role of diagrams within the internal structures and mechanisms operative in the proof system. After Hilbert, proper geometric proofs which employed diagrams could do so only as a cognitive convenience or as an aid to elucidating the textual flow of the argument for the reader. Hence, the backlash against geometric intuition and the rise of pure geometry also provides us with an answer to another question which we posed at the start of this investigation of geometry: why didn't geometry, parallel to logic, ever develop a research program which was concerned with the formally-specified deductive properties of its characteristic representation? At the time of Hilbert, when the requirements of the formalist program caused such metamathematical investigations to become fashionable in logic, the subject matter of geometry had already been mostly absorbed into algebra and analysis. Research in geometry had abandoned traditional Euclidean ruler-and-compass constructions, and the philosophical underpinnings of geometric research had grown remote from the Platonic realism which supported Euclidean geometry. The role of the diagram, associated as it was with the appeal to intuitions and interpretations concerning the behavior of objects in physical space, was at odds with the newer philosophies. Thus, by the time that geometers had the logical tools and conceptual apparatus to have developed purely diagrammatic deductive systems, the role of the diagram had been explicitly limited to that of a simple cognitive aid for a certain class of elementary situations. The general mathematical study of geometry had ceased to be about the sorts of entities which diagrams could reliably or usefully describe.

Part II: Logic

7

Diagrams for Logic

Actual deductive reasoning as it is carried out in the world has multiple aspects: its degree of complexity, the type of representations it employs to represent the information, the quantity of resources it consumes, the persuasiveness it has to the intended audience, the degree of formality it exhibits, the context in which it is embedded, and the extent to which it is correct and can be relied upon, just to name a few. The study of logic over the centuries has tended to concentrate mainly on a single one of these aspects: providing sophisticated and general accounts of correctness for deductive reasoning. In order to build theories in service of this goal, many of the other aspects of deductive reasoning were ignored or restricted in some way. For example, modern first-order logic abstracts away from details of the resource context of reasoning, limits the type of representations employed to finitely long sentences of a formally specified language, forbids the use of probabilities and most other uncertainty metrics, and only admits of tense, belief, and other such modal notions in a limited way. Many of the more recent developments in logic, such as linear logic, involve attempts to relax one or more of these restrictions, and thereby extend the scope of the modern theories of validity to encompass more of the actual phenomena of reasoning as such reasoning is embedded in the world. However, the theoretical limitation of the objects of reasoning to sentences of a language has remained largely unquestioned. With few exceptions, ever since the Greeks we find that formal accounts of generally-applicable deductive reasoning have concerned themselves exclusively with modeling the entailment relations between eternal sentences.

So, even given that the focus of logic has been on the analysis of correctness in reasoning, why has that analysis always reduced to the analysis of valid sequences of sentences in some language? Why haven't other sorts of information-bearing entities been routinely included in the various academic theories and debates about the scope of logic, either as alternate representations for sentences or as the symbolic material for deductive systems in their own right? In particular, why haven't diagrams, whose history stretches back to primitive surveying activities in Babylonian times, and whose usefulness in certain kinds of practical reasoning is undisputed, not been serious candidates for logical investigation until very recently? The answer to this question, which we will give in this part of the book, will require both an examination of some important overall trends in the intellectual history of logic, and an understanding of the individual goals and interests of a few of the dominant figures. But, before beginning, let us first make some general observations about the factors external to the logical theories themselves which nevertheless affect the growth and development of these theories.

One of the most obvious and important of these external factors is that, until very recently, formal systems of reasoning were invented and studied with a specific subject matter in mind. Historically, this subject matter has been distilled from some portion of the actual activity of reasoning as it takes place in the world. For example, Aristotle introduced the syllogism as a "form of words" (*logos*) whose direct application is concerned with (among other things) persuasiveness in the sorts of public disputes in which a scientist could be engaged at the time.[1] The specific characteristic of actual reasoning activity which Aristotle addressed with syllogistic theory was the fact that possessing certain information can sometimes provide legitimate grounds for concluding certain other information, without having to resort to direct observation or experience to verify this other information. For example, if we know that no fish are rational and that all sharks are fish, we can reliably conclude that no sharks are rational, without having to first evaluate the rationality of every shark. Or, to take another example, from the knowledge that all men are rational and that some animals are men, we can conclude solely on the basis of the premises that some animals are rational.[2] However, the phenomenon of valid reasoning is clearly limited in scope. For example, on the basis of the first two pieces of information, we cannot reliably conclude that all fish are sharks. This basic fact about the way that information works in the world – that on certain occasions, certain information can guarantee the truth of a different piece of information – is the fundamental intellectual target which the discipline of logic seeks to

[1] Aristotle, *Prior Analytics* 24b18.

[2] This is an example of the Darii syllogism, and the preceding is an example of Celarent. With Barbara and Ferio, they make up Aristotle's first syllogistic figure.

explain. And, both the direction of logical research and the character of the various logical theories have been driven by the ways in which central metaphysical concepts like "information," "truth," and "guarantee" have been perceived during different periods.

Let us return to the question just posed: why hasn't the use of diagrams in human reasoning activity ever been a serious candidate for logical investigation? My answer to this question will begin by considering the background goals and purposes which logical theories were initially designed to serve, and the sorts of paradigm cases for logical analysis which these goals inspired. By far the most influential of these determinations of the subject matter of logic was the original one, which was primarily due to Aristotle and came about as a result of several factors beyond a simple survey of basic argumentation strategies. As is a familiar pattern in the philosophy of science, the context in which logic was initially developed was guided by the ability of its creators to build on previous work, as well as the ease with which it was possible to integrate the new discipline of logic with other emerging theories, and the overall societal consensus about which areas of investigation would be profitable and useful. Each of these factors pointed at the entailment relations which occur between certain primitive types of propositions as the fundamental reasoning phenomena most in need of explaining, and as the place where a theory of valid entailment could be most clearly applied and tested. This domain, often specialized to disputes revolving around problems in scientific classification, therefore became an important touchstone for the new logical theories.

One of the major reasons for Aristotle's initial choice of this particular subject matter for logic, out of all of the possible types of reasoning behavior, is related to the role he intended logic to play within the context of the large-scale structure of scientific inquiry. For Aristotle, logic was intended to fill a very specific niche in this structure. In his system, the three disciplines of syllogistic, rhetoric, and dialectic comprised the universal arts of argumentation, whose techniques were thought to be fundamental and applicable to knowledge acquisition in all of the theoretical and practical sciences. This view of logic and its place in the intellectual hierarchy was undoubtedly influenced by Plato's earlier success with, and reliance upon, the process of dialectic as the sole tool with which disputes can be legitimately settled. However, we shall see in our investigation that Aristotle's universalist view for logic also placed two very strong constraints on the kind of reasoning phenomena which logic can admit for study. First, Aristotle held that any candidate syllogistic reasoning method must be completely general, and therefore must be sufficiently independent of subject matter that it can be used for valid reasoning in any discipline at all. Second, Aristotle was forced to limit his logic to only describing the subject-predicate relationships which occur in certain basic and general forms of propositions, which he believed were the common ground across the whole of science. These two consequences of Aristotle's insistence on absolute universalism have had extremely important effects on the evolution of logic, with significant consequences for the inclusion of diagram-based reasoning methods in classical logic.

Chapter 8 will therefore concern itself with investigating the early history of syllogistic theory, with the goal of providing an account of why Aristotle structured this theory in the way that he did. As I have just outlined, I will find that Aristotle developed the syllogistic in support of a specific conception about the scope and uses to which logic would be put, and that this conception naturally entails that sentences of a certain limited grammatical form be the principal targets of the theory. This initial demarcation of the proper subject matter of logic, which was based on Aristotle's original account of the range and organization of scientific knowledge, grounded the development of logic from the initial Greek period until the eighteenth century. It also ensured that the logical research of the period was uniformly consistent with the substance-attribute framework of the syllogism, and it secured the position of language as the standard medium in which to express and study valid logical inferences. Considerations of this sort, I will conclude, were responsible for the lack of development of powerful diagrammatic methods in reasoning during the initial Greek period in which the expressive power of the syllogism defined the scope of logic.

In spite of the severe restrictions imposed by syllogistic theory and its underlying metaphysics, however, logical research in the centuries after Aristotle did include some early attempts to represent the various relationships of the syllogism in iconic form. Chapter 9 will therefore take up some of the more significant graphical systems that were developed for the syllogistic. I will devote section 9.2 to a discussion of some of the advantages of using the standard syllogistic formalism to express the reasoning patterns which Aristotle identified as paradigm cases for logic to treat. I will also trace how Aristotle's view about the proper intellectual goals of any logical theory meshed well with the natural patterns of abstraction in a syllogism's form, and thereby reinforced the dominance of language as the preferred medium for investigating logical relationships. Next, section 9.3 will take up some of the earliest graphics which were used to aid the explanation of the syllogism, and analyze the properties of the syllogistic which they stressed. Finally, I will devote section 9.4 to the diagrammatic system of Euler. Euler's work, as we will see, signaled the beginnings of a fundamental change in the basic understanding of the propositions treated by logic, from the classical substance-attribute view based on Aristotle's work, to a more modern class-based approach. This new interpretation of the subject matter of logic also entailed a major change in the ontology of logic, and this provided a natural point at which the question of graphics in logic could be addressed. As we shall see, once propositions are interpreted as claims about inclusion and exclusion relationships which occur between classes, rather than claims about the predication of attributes to substances, certain sorts of graphical mechanisms (based on inclusion and exclusion between figures in the plane) become obvious avenues for investigation. Euler's system used graphics for this purpose, and so we will investigate at length his views toward the system he developed.

Chapter 10 is devoted to the philosophical position of diagrams in symbolic logic. The great changes in logic and the foundations of mathematics during the nineteenth century finally ended the dominance of the syllogism and made possible approaches to the subject matter of logic which were not derived from Aristotelian philosophy. I will use section 10.2 to explore the roots of this change. During the first half of the nineteenth century, the power and generality of algebraic methods in mathematics had inspired the English logician George Boole to create a unified theory of logic which was not based on the classical syllogistic doctrine of substances and attributes, but instead on a more general account of the possible interactions between classes of objects. Boole completely jettisoned Aristotle's original philosophical foundations for the syllogism, and recast logical theory as a set of basic quasi-mathematical operations on uninterpreted classes. His theory is generally regarded as the first work of symbolic logic, and therefore constituted a major turning point in the overall development of logic. For our purposes, the advent of symbolic logic and the subsequent eclipse of the doctrine of the syllogism also had dramatic consequences for the role of diagrams in logic. Three more-or-less distinct intellectual threads can be identified in this change, and I will devote a section of chapter 10 to each of them.

Section 10.3 will discuss the history and development of Venn diagrams for Boolean logic. The initial disappointing reaction of most logicians to Boole's theories explicitly inspired Venn to write his 1881 *Symbolic Logic*, in which he introduced the system of Venn diagrams. Venn's system was the first graphical system to go beyond the expressive and deductive power of syllogistic logic, and he employed the framework of Boolean logic to justify graphical reasoning about problems which could not be addressed syllogistically. This section will explore some of the relevant details of Venn's system, and explicitly contrast it with Euler's. Unfortunately, Venn's initial system was severely limited in its expressive ability and relied on a deductive methodology which was only vaguely defined, and it was not until the subsequent work of Peirce that Venn's original diagrammatic technique was put into the form which is familiar today. Nevertheless, Venn had much to say about the relation of his system to Euler's, and about the relationships between Boolean logic and Aristotelian logic generally, and we shall use this section to examine this.

Sections 10.4 and 10.5 will take up the beginnings of modern mathematical logic in the work of De Morgan, Peirce, Frege, and Hilbert. Their new approaches to logic, while symbolic in nature and grounded in many of the insights of Boole, went beyond pure Boolean logic in several ways, and their success at addressing several long-standing logical issues finally signaled the end of the syllogistic as the sole serious theoretical framework for investigating the phenomenon of valid inference. Importantly, however, the invention of mathematical logic involved more than simply the replacement of one theory of logical consequence with another. The theoretical goals for these new logics were very different than those underlying Aristotle's logic, and because of this, the systems that were inspired by these goals were not able to serve the same philosophical function. The new systems that were developed were both more general (*e.g.*, in the types of entailments they could admit) and more specific (*e.g.*, in their eventual abandonment of the effort to provide an account of coherent thought itself) than the traditional syllogistic ones. Furthermore, the role which these new systems were designed to play in their different conceptions of logical proof forced them to reflect a new and much more limited conception of the scope of logic. The success of the new symbolic logic at filling this role, and its apparent promise as a way to secure the foundations of mathematical truth, caused this new conception of the purpose of logic to itself become dominant over the syllogistic approach. After Hilbert, logic could no longer be conceived (with Aristotle) as a single, universal, and completely general system for describing all possible valid entailment relations which can hold between arbitrary concepts. Rather, logic evolved into the symbolic study of particular sorts of relations between formally-defined structures – relations which were often inspired by, but which are ultimately independent of, any set of intuitively-accepted entailment relations between scientific classifications or simple natural language sentences. Further, the mathematical roots of contemporary symbolic logic ensured that many of the most exciting applications of logic occurred within the foundations of mathematics, and not within the study of argument or reasoning broadly conceived. This fundamental transition, from *logica magna* to *logica utens*, has had important consequences for the possibility of diagrams in logic, and I will conclude Part II by considering them.[3]

[3] These two terms, *logica magna* and *logica utens*, are taken from van Heijenoort 1985. We shall define them more precisely in section 10.5 in our discussion of the influence of Frege and Hilbert on modern logic.

8

The Logical Framework of the Syllogism

The contributions made by the ancient Greeks to the development of logic were enormous. Though it is surely the case that humans have been debating among themselves since the beginnings of language, the first recorded attempts to abstract and study the basic principles governing valid argument were carried out by the Greeks. Aristotle is generally accepted as the person who made the first systematic study of the theory of reasoning and argument, and whose work was the first to define and delimit the study of logic in its own right. He personally staked his claim to originality in the final book of his *Sophistic Refutations*, saying there about his logical investigations:

> Of the present inquiry, on the other hand, it was not the case that part of the work had been thoroughly done before, while part had not. Nothing existed at all... [O]n the subject of deduction we had absolutely nothing else of an earlier date to mention, but were kept at work for a long time in experimental researches.[1]

Let us call this claim Aristotle's *originality* claim. Evaluating the originality claim will illustrate some important features about Aristotle's conception of deduction. For, although Aristotle's writings on the syllogism in the *Prior Analytics* are the first known attempts to systematically address the general structure of the entailment relations used in dialectic and demonstration, it is

[1] Aristotle, *Sophistic Refutations*, 183b34.

certainly *not* the case that precise reasoning was completely unknown and unstudied before Aristotle. Prior to Aristotle's work on logic, there was important foundational work in mathematics, physics, and rhetoric, aimed at clarifying the standards of expression and demonstration in their respective disciplines. Because of this, the reasons that Aristotle discounted these efforts in his originality claim will tell us a great deal about the intended subject matter of the new discipline he was defining. We will concentrate on the case of geometric reasoning, since we have addressed it at length in Part 1 of the book, and since Aristotle was clearly influenced by its example.

Let us briefly review the situation in geometry at the time of Aristotle. Although Aristotle died before the time of Euclid's work, his logical research had the benefit of a roughly two hundred year history of work in abstract geometrical reasoning, often from an explicit set of first principles. The classical commentator Proclus gives credit for proving the first geometric theorem to Thales in the sixth century BC.[2] However, the Pythagoreans, active between about 585-400BC, were probably the first to embrace the concept that geometry could be studied as a deductive, abstract science, rather than as a science whose precepts and methods should be determined empirically. Several of the great classical geometers who lived before Aristotle, including Hippocrates and Eudoxus, were inspired by the work of the Pythagoreans, and as a result of their successes, the tradition of presenting geometry as a deductive system from a set of axioms and definitions was well-established by Aristotle's time. Aristotle was undoubtedly familiar with this approach to geometry, as he uses geometric examples several times in his writings. However, he never attempts to justify even a simple geometric theorem by reference to syllogistic patterns of reasoning. Coupling this with Aristotle's claim that no part of his logical work had been done before, we can see that he must have viewed his logic as distinct in subject matter from, and addressing a quite different sort of problems than, the growing edifice of geometry.

What is the essential difference that Aristotle saw between syllogistic reasoning and geometric reasoning, which led him to claim that there was no other prior work on deduction which was worthy of mention? Aristotle's definition of a deduction is very general, going substantially beyond the proof-oriented definitions common in formal logic today:

---

[2] See Kneale and Kneale 1962, pg. 3. Thales' priority in geometric proof is disputed in Kline 1972, pg. 28.

> Now, a deduction is an argument in which, certain things being laid down, something other than these necessarily comes about through them.[3]

This definition would certainly encompass the sort of reasoning found in Greek geometric practice, as well as many other sorts of arguments which cannot be put into strict syllogistic form. Hence, Aristotle cannot have held *simpliciter* that geometric reasoning does not involve deductions. What position is he articulating in the originality claim, then? We can find an interesting but obscure clue in the *Topics*, where Aristotle refers to geometry as a "special science" and observes that deductions in it may exhibit characteristic fallacies based on the unique methods which geometry employs.

> Further, besides all the deductions we have mentioned there are the fallacies that start from the premises peculiar to the special sciences, as happens (for example) in the case of geometry and its sister sciences. For this form of reasoning appears to differ from the deductions mentioned above; the man who draws a false figure reasons from things that are neither true and primitive, nor reputable. For he does not fall within the definition; he does not assume opinions that are received either by everyone or by the majority or by the wise ... but he conducts his deduction upon assumptions which, though appropriate to the science in question, are not true; for he effects his fallacy either by describing the semicircles wrongly or by drawing certain lines in a way in which they should not be drawn.[4]

In this passage Aristotle is worried about the status of fallacious reasoning based on "false figures." Presumably, he is thinking here about geometric diagrams which depict impossible situations or which do not accurately track the proof being developed. Aristotle had earlier distinguished two subkinds of valid deduction based on the epistemic character of their premises: demonstrative and dialectical. Demonstrative deductions involve reasoning from premises which are true and primitive; dialectical deductions involve reasoning from less-well-grounded reputable opinion. He also identified as "contentious" those deductions which either proceed from premises which appear reputable but are not, or those which contain a logical error in their reasoning. But, as Aristotle writes, creating a false figure is different in its "form of reasoning" from the fallacy exhibited in each of these sorts of contentious deductions. What does he intend here? It seems that the method used in the reasoning is at the crux of Aristotle's problem: drawing

---

[3] Aristotle, *Topics*, 100a25.
[4] Aristotle, *Topics*, 101a5.

lines erroneously or creating incorrect semicircles ("effecting the fallacy") is not the same as performing valid deductions on disreputable opinions, nor does it involve a syllogistic mistake. Rather, the construction methods which are used in geometric reasoning are specifically tailored to the form of the premises, which are "peculiar" in special sciences like geometry, and the scope of their standards of correctness is limited to geometry. Hence, Aristotle classifies fallacies arising from these methods differently from syllogistic reasoning which exhibits a mistake, or is based on opinions which are only held by the foolish. And, importantly for our purposes, this passage suggests that Aristotle saw a distinction between the form of reasoning which takes place in geometric constructions, and the form of reasoning which can be described by the syllogism.

This interpretation fits well with what we know about Aristotle's general view about reasoning and knowledge acquisition in geometry and its "sister sciences." As his geometry example in the *Topics* implies, the processes of discovery which are characteristic of these sciences may involve methods which go beyond purely syllogistic reasoning based on true premises or reputable opinion. These methods may involve, for example, geometric reasoning based on figures, *reductio* reasoning involving the explicit assumption of false premises, or inductive discovery of primitive propositions based on experience. Further, these methods are typically quite specialized in their application: the sorts of ruler-and-compass construction techniques which were indispensable to discovery in classical geometry, for example, were not appropriate to investigation in physics or medicine. Aristotle's philosophy of science distinguished the discipline-local methods appropriate to inquiry in the individual sciences from the universal arts of argumentation, which are applicable to reasoning and discovery across all the disciplines. The study of these universal arts – dialectic, rhetoric, and analytic (syllogistic) – were for Aristotle the province of the philosopher, rather than the practitioner of any individual science:

> Plainly, therefore, it falls to the philosopher, *i.e.* the student of what is characteristic of all substance, also to investigate the principles of trains of reasoning.[5]

The sort of diagram-based reasoning found in geometry would not, therefore, be an appropriate topic of investigation for the philosopher, because

[5] Aristotle, *Metaphysics*, 1005b7.

(*contra* syllogistic deduction) its techniques are not general enough to be applied to any domain.

From the preceding discussion, then, we can see how Aristotle's division of general investigative procedure into that which was specific to a science and that which was universal to all sciences provides a framework for understanding his originality claim in the *Sophistic Refutations*. Certainly Aristotle was aware that a developed tradition of exact reasoning existed in many sciences, including geometry, prior to his work in logic. This work may even have qualified as a deductively-based method in Aristotle's broadest sense of deduction, but Aristotle was not referring to that type of deduction when he claimed "on the subject of deduction we had absolutely nothing else of an earlier date to mention." Rather, he was referring to his work laying out the framework of the universal art of deduction: general reasoning applicable to every study of substance. The figure-based methods of geometry, just like the numerically-based methods of mathematics, are not applicable to *any* study of substance, and so they would not count as a precursor to Aristotle's study of syllogistic, and would not be a counterexample to the originality claim.

At this point, let us pause from our analysis of Aristotle's originality claim, and consider its implications for our larger question about the lack of diagrammatic techniques in logic. Aristotle distinguished the universal arts of argumentation from the methods of all the other sciences by the generality of their application. As it happened, the study of one of these arts, the analytic, became fairly rapidly separated from dialectic and rhetoric, and its syllogistic technique became the foundation of the Aristotelian approach to logic.[6] This means that a certain sort of universalism was explicitly present at the founding of logic, and more importantly, was an essential feature of the way the discipline was defined. The prime area of application for precise diagrammatic reasoning techniques – geometry – would not have been considered an appropriate area for logical research by Aristotle, not because of its graphical nature *per se*, but because its methods were insufficiently general for the discipline he had in mind. This fact alone may provide much of the explanation for why diagram-based reasoning of any sort has never been considered a topic of investigation in traditional Aristotelian logic.

---

[6] McKeon's introduction to his Aristotle collection provides us with a potential theoretical reason for the gradual narrowing of the subject matter of logic to the study of the syllogism: "Dialectic and rhetoric are methods as well as arts because their use depends on knowledge of opinions, the opinions of all or most men, or the opinions of groups of men envisaged as the particular audiences or recipients of forms of communication. The two *Analytics* are not methods, because the only subject-matter knowledge they are concerned with is stated in propositions warranted by the sciences among which they construct and examine inferential relations." See McKeon 1973, pg. 2.

Aristotle's explicit universalism about the subject matter of logic has had another, deeper effect on the subsequent development of the field. By straightforwardly enshrining general applicability as the defining characteristic of the reasoning which was to be studied by logic, Aristotle limited logic's subject matter to those argument types which could be applied to the study of all substance. However, in doing this, he also rejected the notion that it is the task of the logician to study valid reasoning *regardless* of its form or degree of generality. This core Aristotelian principle – that logic should address only universally applicable patterns of reasoning – certainly has contributed to the lack of attention paid to diagrammatic forms of inference. More importantly, though, it has also been a powerful underlying force on the overall evolution of logic as a discipline. As a result of this principle, logicians working in the tradition of Aristotle have typically refrained from investigating many of the individual reasoning patterns which are found in geometry, physics, medicine, or a host of other sciences, because these patterns are not universal enough to be applied to an arbitrary subject matter, and therefore would not qualify as truly "logical" in nature. For this reason, Aristotle's exact reasons for his choice of disciplinary boundaries for logic would be an important topic in any comprehensive history of logic. However, space and focus considerations prevent us from investigating these reasons any more fully here.[7]

Returning to the more narrow subject matter of the syllogism, the evidence of Aristotle's universalist goals for the analytic art is present in ways which go beyond the syllogism's assigned place in the hierarchy of scientific techniques. We can find this evidence in the way in which Aristotle integrated the theory of the syllogism into classical accounts of predication and the structure of knowledge. Again, let us use the originality claim as a lens with which to view this issue. Even given Aristotle's distinction between the universal arts and inquiry-specific methods, it seems that Greek progress in understanding the (other) universal art of dialectic would be quite relevant to his study of the syllogism. And here, Aristotle's originality claim seems much more difficult to defend. Public disputation according to a set of rules had been practiced since substantially before Aristotle's time; indeed, his *Sophistic Refutations* is primarily a handbook of strategies for par-

[7] We should also note that disciplinary boundaries are the sort of thing whose establishment is rarely the result of calmly reasoned arguments from first principles, and so it would be unusual if this one could be successfully defended through such arguments. The precise reasons for Aristotle's generality constraint in logic are probably as much a result of his own reaction to preexisting divisions within Greek philosophy as anything else. Because of this, an adequate investigation of these reasons would involve a far more detailed examination of Aristotelian philosophy than we can afford here.

ticipants in these debates to employ against each other. These disputes often employed the process of dialectic – a *reductio* strategy of showing by step-wise argument that the position taken by one's opponent results in a false or contradictory statement, and then concluding that the original position could not have been the case. Evidence for the respect given to dialectic in Aristotle's world can be found in Plato's *Republic*, where Socrates both links it with the practice of philosophy and worries about its seductiveness to the youth:

> For I fancy you have not failed to observe that lads, when they first get a taste of disputation, misuse it as a form of sport, always employing it contentiously.... And, when they have themselves confuted many and been confuted by many, they quickly fall into a violent distrust of all that they formerly held true, and the outcome is that they themselves and the whole business of philosophy are discredited with other men.[8]

By the time Aristotle wrote the *Prior Analytics*, the long popularity of public dispute and the use of dialectical methods had given rise to a considerable body of knowledge concerning argumentation theory. We can find several examples with which Aristotle, as Plato's student and the foremost philosopher of his time, would surely have been familiar. The early Sophists had developed rudimentary classifications of the different types of sentences. Plato, in the *Sophist*, examined the structure of simple statements, and classified their constituents into what we would now call noun phrases and verb phrases. In the *Euthydemus*, Plato showed that validity does not attach to particular configuration of words, but to the propositions to which they refer. And, Kneale and Kneale 1962 discuss a fragmentary early fourth-century BC text, the *Dissoi Logoi*, in which the author takes up basic questions of the nature of truth and falsity, and relates them to a set of antimonies.[9] The sophistication of the work in this tradition casts doubt on Aristotle's claim he could find nothing in it that was relevant to his logical research. Many of the basic accomplishments of the study of dialectic, such as the analysis of sentences into subject-predicate form, the understanding of predication, and the Platonic argument for the coherence of negation, are implicit in the very foundations of the syllogism.[10] Aristotle was certainly

---

[8] Plato, *Republic*, 539b.

[9] Kneale and Kneale 1962, pg. 16.

[10] Problems with supposing the negation of something that is known to be the case had exercised the Greeks since Parmenidies. Plato's *Theaetetus* and *Sophist* include arguments which address these basic logical issues. See Kneale and Kneale 1962, pg. 21.

aware of these accomplishments, and his work bears their imprint. In particular, these basic assumptions concerning the structure and form of knowledge and its expression are highly relevant to the question of why Aristotle chose to define the subject matter of logic as he did, and concentrate it exclusively on a certain restricted sort of linguistic argumentation.

In order to explain this, let us turn to Aristotle's explicit definition of the syllogism and its role. When Aristotle introduced the syllogism in the *Prior Analytics*, he wrote that:

> A syllogism is a form of words in which certain things are assumed and there is something other than what was assumed which necessarily follows from things' being so. By 'from' I mean 'because of'; and 'follows because of things' being so' means that no further proposition is needed to make the 'following' necessary.[11]

This does not, on its face, give any specific theoretical role or intended subject matter for the syllogism. However, in another passage, Aristotle left no doubt that his expected use for the syllogism was comprehensive:

> We must now state that not only dialectical and demonstrative syllogisms are formed by means of the aforesaid [syllogistic] figures, but also rhetorical syllogisms and *in general any form of persuasion*, however it may be presented. For every belief comes either through syllogism or from induction.[12]

This passage, although overstated, confirms that syllogistic reasoning was central in Aristotle's picture of the structure of knowledge.[13] In the *Posterior Analytics* and the *Metaphysics*, Aristotle provides more detail. He describes how he conceives of a science as starting with a body of primitive propositions concerning substances and their various types of causes, and a collection of assertions which link them to other, less fundamental statements of the science.[14] The syllogism, he believed, was an important tool

[11] Aristotle, *Prior Analytics*, 24b18. Note the similarity to the definition of deduction given in *Topics* 100a25 and quoted earlier.

[12] Aristotle, *Prior Analytics*, 68b9, emphasis added.

[13] Aristotle also refers several times to arguments which are valid but not syllogistic. One example of this would be reasoning from conditional premises, such as *modus ponens* and *modus tollens* arguments. Also see, for example, the discussion at *Prior Analytics* 47a.

[14] In this discussion, I will use "cause" in its broad Aristotelian sense as, roughly, a principle which licenses the ascription of an attribute to a substance.

for the extension and systematization of general scientific knowledge in any discipline. More specifically, he thought that deduction via the syllogistic forms would be a reliable way to expose the law-like universal reasons which undergird particular facts and observations, and thereby facilitate arriving at something he later called a "demonstrative understanding." For example, we find Aristotle saying:

> For here it is for the empirical scientists to know the fact and for the mathematical [those scientists proceeding via logic] to know *the reason why*; for the latter have *demonstrations of the explanations*, and often they do not know the fact, just as those who consider the universal often do not know some of the particulars through lack of observation.[15]

This passage, and others like it, illustrates how Aristotle thought of syllogistic as a critical device with which to understand the experiential patterns which arise from the operation of (universally applicable) natural laws.[16] Science, for Aristotle, was concerned with the discovery of various types of causes and their operation upon substance. Once the causes have been discovered, syllogistic methods can be used to state and validate the relations that these causes bear to the substance; indeed, the *Posterior Analytics* is mainly devoted to the question of the selection of the most appropriate syllogistic middle term when reasoning demonstratively or dialectically in scientific inquiry. These relations can be expressed and known via the universals referred to in the quote above, and according to Aristotle, it is the knowledge of these universals which allows us to transcend bare experience and forms the basis of the knowledge of the scientist.

How is Aristotle's theory of scientific knowledge relevant to our inquiry? The reason has to do with the doctrine laid out in Aristotle's *Categories*, which addresses the basic nature of predication between terms. According to the *Categories*, the ultimate subject of predication and constituent of knowledge is what Aristotle calls primary substance:

> Thus everything except primary substances is either predicated of primary substances, or is present in them, and if these last [primary substances] did not exist, it would be impossible for anything else to exist.[17]

---

[15] Aristotle, *Posterior Analytics*, 79a3, emphasis added.

[16] This use of the syllogism is the locus of one important difference between Aristotle and Plato. For Plato, the ultimate tool for the extension of knowledge and the resolution of contradictions is the process of dialectic.

[17] Aristotle, *Categories* 2b4.

This means, as Kneale and Kneale 1962 points out, that the basic truths of the sciences will all be of the form, "this [primary substance] is (or is not) such-and-such," and any other scientific statement, especially the aforementioned universals, will be logically dependent upon a collection of prior truths of this form about primary substance.[18] Further, Aristotle thought that these basic truths can vary only in certain well-defined ways (the choice of substance and attribute, the presence of a quantifier-term, the presence of negation, or the presence of a modality-term), and he held that all predicates could range over every possible subject:

> Thus, we assume that every predicate can be either truly affirmed or truly denied of any subject.[19]

Therefore, by abstracting from particular subject matter in order to study the art of the analytic, Aristotle was able to ignore variation in the substance and property terms, and by temporarily setting aside modality considerations he was able to arrive at the four primitive classes of categorical proposition described in *On Interpretation*: the particular affirmative (Some S is P), the particular denial (Some S is not P), the universal affirmative (Every S is P), and the universal denial (No S is P). Aristotle then further observed that these four propositional types could be subdivided into sets of contradictories and contraries, and that this division gave rise to entailment relations between them. He combined all of this reasoning to structure these four proposition types into the basic elements of the first-figure syllogism in the *Prior Analytics*. Indeed, Aristotle claims that these four types of statement are the fundamental building blocks of all types of demonstrations:

> It is necessary that every demonstration and every syllogism should prove either that something belongs or that it does not, and this either universally or in part, and further either ostensively or hypothetically.[20]

---

[18] Kneale and Kneale 1962, pg. 31.

[19] Aristotle, *Posterior Analytics*, 71a13.

[20] Aristotle, *Prior Analytics* 40b23. Strictly speaking, Aristotle is here only claiming that the *conclusion* of a syllogism has to be in this form, although the development of the syllogism in the *Prior Analytics* makes it clear that he assumes that the premises need to be in this form as well.

Given Aristotle's earlier claims about the comprehensiveness of syllogistic reasoning, it is clear what a strong claim he is making about the fundamental nature of precise reasoning. For Aristotle, the syllogism is designed to be a tool to discover and verify truths, and all of these are dependent on truths of a certain basic form, and therefore the syllogism should be structured around reasoning with these sorts of basic truths.

From the preceding, we can now see how Aristotle's account of knowledge as consisting of basic truths structured in a particular constrained subject-predicate form, coupled with the requirement that the deductive art be independent of any specific subject matter, led him to propose the syllogism in the particular sentential format that is still familiar today. And, apropos our question about the lack of attention paid to diagrammatic reasoning, we can also see how Aristotle's view of the fundamental form and structure of knowledge implicitly guided the further advancement of logic to emphasize entailments between propositions of this type. Under Aristotle's framework, any differences between propositions which could not to be ascribed to variation in their logical form, such as issues of indexicality or the presence of relations, could usually be accounted for by appeal to specific characteristics of individual predicates, and hence would not be relevant to the topic of logic. Instead, the emphasis in logic on the study of the patterns of valid predication of categories to substances, or more generally of properties to subjects, dominated the field for almost two thousand years, from Aristotle's time until the Renaissance, and did not fully disappear until the advent of modern logic at the beginning of the twentieth century.

In conclusion, let us review what we have accomplished so far in this chapter, and tie it explicitly to our question of why diagrammatic reasoning techniques have been largely ignored in formal work in logic. We have identified two powerful underlying reasons why Aristotle would have ignored graphically-based methods in his initial theorizing about the nature of reasoning phenomena. The first is related to the universality which he required of the reasoning techniques to be addressed by his theory. In order for syllogistic reasoning to be capable of the generality which Aristotle intended, and thus able to be uniformly applied across all possible types of inquiry, its techniques must themselves be absolutely free of any dependency on the peculiarity of an individual subject matter.[21] The chief Greek examples of reasoning which involved a strong graphical component – reasoning in geometry, cartography, and the like – were all strongly tied to their particular subject matter, and so would qualify as discipline-specific methods in Aristotle's scheme. For example, the ability to construct the per-

[21] Recall our discussion in Part I about how these sorts of universalist goals came to be present in geometry through the work of Pasch and Hilbert.

pendicular bisector of a given line segment is an important technique in elementary geometry, but has no application to reasoning about the genus and species relationships in biology. Therefore, although he never addressed the point explicitly, we should not be surprised that Aristotle did not attempt to give an account of universally valid reasoning using graphical or diagram-based techniques – he had no reason to believe that they could exhibit the required generality, and many examples that suggested that they could not.

In addition to this issue about the dependence on subject matter shown by existing diagrammatic reasoning techniques, we have identified a second underlying reason why Aristotle would not have included such techniques in his development of the syllogism. Aristotle's theory about the underlying form of scientific knowledge restricted him to considering propositions of a particular, limited sort as the fundamental building blocks of valid reasoning. Recall his earlier statement on this subject: "it is necessary that ... every syllogism should prove either that something belongs or that it does not, and this either universally or in part." Given this view of the basic determinants of knowledge, the notation he devised for describing the admissible patterns of valid general reasoning – a pattern of sentential schemata with variables standing in place of the subject and predicate – naturally brings out the relevant features. Consider, for example, his formulation of Celarent: "if every S is M and no M is P, then no S is P." The non-variable terms in the statement of this syllogism serve to neatly identify the combination of the logical attributes which Aristotle recognized in his discussion of the nature of predication. Further, this linguistically-oriented way of constructing his theory of general validity in reasoning meshed naturally with the basic phenomena he intended to describe – the entailment relationships which hold between truths of a particular form – as well as linked with the previous work on rhetoric that Plato and the Sophists had done. So, given his view of the ultimate targets of his theory, it is unsurprising that Aristotle built it in the way that he did.

We have now identified two reasons why Aristotle did not allow for diagrammatic reasoning techniques in his theory of the syllogism. In the following chapters, we will shift to investigating how these reasons played out in the history of logic after Aristotle. However, we should note that from our modern perspective, and even granting Aristotle's classical epistemology and philosophy of science, these reasons are not compelling. Against the first reason, Aristotle never gave a broad-based argument which explained why the syllogism as he formulated it was the preferred way to structure a theory that addressed generally-applicable reasoning. We have speculated that he ignored the possibility of graphical techniques in reasoning because he was influenced by the example of Greek geometry, but this example of a successful domain-specific graphical technique does not constitute an argument for the impossibility of similar general techniques. And against the second, Aristotle never argued that the requisite reasoning on basic truths had to be done strictly by using a sententially-based notation like the one in which he formulates the syllogism, rather than with (at least in part) some suitably structured set of graphical tools. Given these two problems, then, our strategy in the following chapter will be twofold. First, we will reexamine some of the technical features of Aristotle's syllogistic doctrine which contributed to its complete dominance as a theory of reasoning in the two thousand years following Aristotle's death. Second, we will examine how Euler, the inventor of the first serious graphical technique for syllogistic logic, perceived his creation. What did he view himself as doing, and in what ways did he see his graphical system as an improvement on Aristotle's foundation?

9

Diagrams for Syllogistic Logic

9.1 Introduction

The emergence of the first serious graphical techniques in logic was made possible by several changes in the philosophical background in which the syllogistic was embedded. The most prominent of these changes was a fundamental shift in the way in which the propositions dealt with by Aristotelian logic were viewed. Let us briefly review the classical picture described in the previous chapter. Recall that for Aristotle, the ultimate subject matter of logic consisted of abstract propositions of a particular limited form, which described whether or not a given substance possessed a certain attribute, or universal. The proposition's subject could be quantified in simple ways, and certain kinds of negation were possible. The familiar schematic sentences of the syllogistic figures are reflective of this abstract form. Aristotle's choice of these types of sentences as the foundation of syllogistic theory was rooted in two things: his view that the ultimate structure of scientific knowledge was based on facts about the characteristics of primary substances which could be completely described by propositions of this form, and his view that all of the relevant explanatory connections at the heart of each individual discipline could be discovered and validated by constructing syllogistic trains which start with these propositions. Aristotle's central claim for his logic, that it would be applicable to valid reasoning across all types of scientific inquiry, was rooted in these two distinct characteristics of it.

Let us give names to these two characteristics of classical Aristotelian logic, and examine them a bit more closely. First, his theory of the syllogism offered *expressive completeness*. By this, I mean that the fundamental components of his theory possessed sufficient power to provide unique representations for all of the basic propositions which were relevant to the accepted subject matter for logic. Aristotle believed that the schematic sentences of his logic, through the mechanism of variable instantiation, would be guaranteed to capture all of the primitive propositions of a science. This was a consequence of the just-discussed doctrine of the *Categories* about the epistemic primacy of the predication of attributes to primary substance, plus the presumed ability of his language to express all of the relevant concepts and logical relations. Second, though, the reasoning component of syllogistic theory also offered *deductive completeness*. By this, I mean that in addition to simply being able to express the relevant propositions, his theory also included derivation mechanisms which were powerful enough to ensure that any entailment between facts in the intended subject matter had a correlated parallel deduction in the logical theory. Because the various syllogistic forms (through Aristotle's combinatorial exploration of the various syllogistic figures and moods) encompassed all the possible valid relationships between categorical sentences, its logical power was guaranteed to be sufficient for Aristotle's intended target of demonstrative reasoning.[1] These two features of syllogistic theory combined to yield a powerful and successful theory of reasoning, and although they were certainly not always explicitly recognized, their combination provided a powerful bulwark against any competing theory. Further, the foundational Aristotelian doctrines which ensured these two types of completeness, especially that of the fundamental subject-predicate structure of the propositions treated by logic, became some of the underlying dogma of the classical approach to logic.

Because Aristotle did not have anything like a formal theory of semantics, however, we should note that our usage of "expressive completeness" and "deductive completeness" amounts to a generalization of the modern notions. At base, it is clear that both of these kinds of completeness must be specified relative to some intended subject matter for the logic, against which the notion of valid entailment can be defined. Today we have explicit semantic theories of the mathematical structure of these domains, and because of this we can precisely say whether, for example, a given system of propositional logic is either expressively and deductively complete relative to a standard truth-table semantics. In this book, when we evaluate the scope and adequacy of Aristotle's syllogistic theory and subsequent systems

[1] In his writings on the syllogistic, Aristotle occasionally hedged on whether or not he thought that it would encompass the entire common fragment of reasoning practice across the sciences. As we have seen earlier, he both writes that all valid reasoning is syllogistic, and refers to seemingly general but non-syllogistic reasoning (see, *e.g.*, *Prior Analytics* 47a24 ff). Perhaps he hoped that further development of his theory would eventually bring these other reasoning forms, such as conditional reasoning with *modus ponens*, under the umbrella of his theory.

like Euler's and Peirce's, we will be making use of the same core notions. However, our particular usage of these terms will of necessity not be as exact as the modern practice, because logicians prior to the late nineteenth century were not precise when specifying the relationship between the logical framework which they proposed and the external reasoning domain which they wanted to describe. Additionally, as views about the structure of this external domain for logic evolved over the centuries, the context in which we will apply the notions of expressive and deductive completeness to different logical systems will evolve as well. Consequently it is only possible to use general notions when attempting to directly compare theories originating in different logical periods. We will first see this in our examination of the work of Euler in section 9.3.

An important task of chapter 9 will be to document how certain details of Aristotle's view of the basic fabric of knowledge were largely incompatible with the familiar circle-based attempts at diagrammatic representational strategies in logic. For this reason, the graphical ideas of Euler did not emerge until the rise of a new, more general view of the structure of the logical proposition based on interactions between the extensions of class terms. But, before describing how these changes occurred, it will be helpful to first use a section to detour from our main line of argument and briefly review some of the background technical features which were inherent in Aristotle's particular sentential formulation of syllogistic theory. We will use - section 9.2 to concentrate on the way in which his account of the syllogism handled abstraction, because this feature was critical to the ability of his theory to claim both expressive completeness and deductive completeness over its intended subject matter. Then, the following two sections will use this background material in order to illuminate their discussion of the evolution of graphical notations for classical logic. First, in section 9.3, we will examine the various Greek and medieval graphical notations which were used in service of the syllogistic. Then, section 9.4 will take up Euler diagrams. We will find that the symbolism used in Euler's diagrammatic system was only possible because of a significant theoretical change in the ontology of logic – the use of the extensional interpretation of properties as a critical part of the theory of deductive systems – and that this change of view toward the salient features of logic's subject matter was what allowed the use of intersecting regions to represent the sort of mediate reasoning described by the syllogistic.

Also, for the remainder of our investigation of diagrams in logic, it will be important to distinguish a particular piece of syllogistic reasoning from the abstract theory which describes it. Henceforth, the term "syllogism" will be used to refer to a completely-instantiated concrete piece of valid reasoning, with specific major, minor, and middle terms. The "syllogistic" will refer to the theory of valid reasoning first conceived by Aristotle, which revolves around a group of schematic sentences used to classify different individual syllogisms.

9.2 The Linguistic Formulation of the Syllogism

An interesting way to introduce our discussion about abstraction in syllogistic theory is to consider the following question: must the abstract information flows described in the syllogistic necessarily be represented in a linguistic, sentential format, or is it only Aristotle's particular representation of this type of reasoning in a standard form which is sentential? From our modern perspective, we can see that the first possibility cannot be the case. Certainly, the primitive propositions which Aristotle intended to capture in the syllogistic can be conveniently expressed within a certain sort of restricted natural language framework, and this framework also nicely decomposes the logical structure of these abstract propositions into individual sentential particles. However, we now know that these relationships can also be expressed by any sufficiently discriminating system of physical marks, whether sentential or graphical in format, through the use of a suitable evaluation function linking the marks to an underlying semantic model. Interestingly, even before the framework of modern semantics was developed, logicians had wondered if the syllogistic's connection to natural language expressions simply resulted from a coincidence: that the abstraction patterns required by the theory just happened to mesh nicely with the grammatical features provided by the language in which Aristotle was working. For example, Venn referred to this issue when he wrote:

> Some philologists have recently directed an attack against the whole science of Formal, viz. Aristotelian or Scholastic Logic, on the ground that it is largely made up of merely grammatical necessities or conventions; nearly all its rules being more or less determined by characteristics peculiar to the Aryan languages. Thus Mr. Sweet ... says that as a consequence of philological analysis, "the conversion of propositions, the figures, and with them the whole fabric of Formal Logic falls to the ground."[2]

[2] See Venn 1894, pg. xxvi fn. 1. All citations are to the 2nd edition, although for the purposes of placing Venn's work in nineteenth century mathematical developments I will use 1881, the year of the first edition.

Whether or not this is an accurate account of the dependence of classical logic upon the structures of natural languages in which they are formulated, it is clear from our modern point of view that it is only Aristotle's *notation* for the syllogistic which is essentially language-based.

This is not to say, however, that Aristotle's particular language-based formulation of the syllogistic patterns was a mere notational convenience. On the contrary, it was intimately connected with the ability of his theory to support the two kinds of completeness claims mentioned above. With regard to the theory's deductive completeness, the formal ability of any theory to capture valid reasoning is intertwined with the ability of its notation to support the kinds of transformations required by the logical relations which occur in the underlying subject matter. Language suited this purpose well. And, with regard to expressive completeness, Aristotle's examples continually show that he viewed reasoning as it occurs in language as the most important touchstone for his theory, and so it was important that the theory's machinery would be applicable to this domain with relative ease.[3] For these two reasons, the medium of natural language would have been both obvious and ideal to Aristotle for the presentation of the valid syllogistic forms of reasoning.

However, language also carried many other advantages as a theoretical substrate: it was compact, expressive, already familiar to his audience, possessed of an appropriate degree of specificity, and relatively well investigated by earlier philosophers. Further, the attractiveness of syllogistic theory to other philosophers was undoubtedly linked to some features of his specific language-based formulation of the key theoretical constructs: the convenient repetition of the major, minor, and middle terms; the clear syntactic parallels occurring between the three sentences of a syllogism; the method of using variables ranging over types of terms to represent whole groups of sentences; and the consequent ability to directly instantiate syllogistic forms as specific syllogisms in a target discipline.[4] In order to be accepted, any alternate notation for logic would have to compete with all of these advantages.

Let us examine in more detail Aristotle's use of variables in the presentation of his theory. Although he typically used concrete examples to

[3] Indeed, when Aristotle gives the details of the theory, he seems to retreat from the broad definition cited earlier from the *Topics*, and concern himself mostly with arguments which consist of exactly two premises and a conclusion, where the terms of the conclusion are related in the premises through an intermediate, or middle, term. Because of this, syllogistic reasoning is often referred to as *mediate* inference. See, *e.g.*, *Prior Analytics* 41b36.

[4] I am assuming here that the translations I have of Aristotle's work, and the presentation of the theory of the syllogism contained therein, mirror the capabilities Aristotle had in the original Greek .

illustrate the different possible syllogisms, he was quite aware that any individual syllogism would be an instance of an abstract reasoning pattern. That is, when interpreted in the light of his theory, individual syllogisms could be seen to contain regularities which were characteristic of universally valid reasoning. Describing and characterizing these regular patterns, as opposed to the specific deductions to which they gave rise, were at the core of his theory, and formed the basis of his claims about the unlimited applicability of his logic. The use of variables in natural language sentences, with the goal of creating a metalanguage for these patterns, was critical to this effort. Consider Aristotle's initial description of what we would now call the Barbara and Celarent syllogisms:

> If A is predicated of every B and B of every C, A must be predicated of every C, by our previous account of *predicated of all*. Similarly if A is predicated of no B and B of every C, A will apply to no C.[5]

His use of letters to stand for terms, and his subsequent suggestion of different example terms to substitute for the letters, reinforce that he intends to refer to a particular inference *schema*. His claim was that the various referring expressions denoted by the letters A, B, and C – the major term, the middle term, and the minor term – could be instantiated with any terms of the correct sort from the language, and the resulting syllogism will be a valid, though possibly not sound, piece of reasoning. With the use of this replacement scheme, Aristotle invented the now-familiar method of using metalinguistic variables to refer to large classes of individual expressions which share a common form.

Although to the modern eye this technique appears obvious and routine, it is worth remembering that at the time the use of variables with specific replacement rules was a major conceptual breakthrough. The great achievement of Aristotle's logical work was to successfully raise the study of entailment from the study of individual deductions to the study of these reasoning patterns contained within them. Kneale and Kneale 1962 suggests that in earlier writings,

> [the required] generality is indicated by a rather clumsy use of pronouns or by examples in which it is left to the reader to see the irrelevance of the special material."[6]

---

[5] Aristotle, *Prior Analytics* 25b37. Emphasis is in original.

[6] Kneale and Kneale, 1962, pg. 61.

By introducing a compact method in which the logical form of many different specific arguments could be distilled into a single notation, and by using this notation as the foundation of his deductive theory, Aristotle succeeded in formally refocusing logic away from rhetorical issues of whether individual arguments were convincing or not to an audience. Aristotle himself was probably not aware of the magnitude of this change; for example, in chapter 1 of the *Topics*, he is still worried about the status of arguments which are acceptable to all men, versus those acceptable only to the wise. However, because of the breadth of the possible substitutions for variables, and the clearly unsound deductions which could result, it rapidly became clear that the defining issue of the nascent discipline of logic was not the soundness or persuasiveness of arguments, but their validity. Thus, we can see that Aristotle's invention of variables, and the consequent emphasis of his theory on the form of arguments, was critical to the birth of the modern discipline of logic. When combined with Aristotle's previously mentioned universalism toward the class of reasoning which logic should address, the invention of the variable established that the logic's fundamental property of study would be the *validity* of the most general sorts of arguments.

Retreating to the standpoint of evaluating Aristotle's theory *qua* theory, however, we can see that his use of this technique provided him with three very important capabilities. First, it gave Aristotle a precise and compact way of expressing his syllogistic principles, in such a way that it was clear (given a prior division of terms into categories) whether or not a particular piece of sentential reasoning conformed to the principle. This method abstracted the principles involved from the details of different examples, and enabled both Aristotle and his commentators to focus cleanly on the logical principles at stake. Second, it gave Aristotle a principled way to structure and impose a hierarchy of importance on his syllogisms. Because, for example, he was able to use these schemata to derive the second- and third-figure forms from the first-figure ones, he was able to propose that certain syllogisms were conceptually more fundamental than others. Third, and most importantly for our purposes, his use of variables for the predicate and the object of predication in the syllogistic theory guaranteed that his theory could be universally applied to all specific subject matters – that it was expressively complete. Given Aristotle's conception of the basic subject-predicate form of scientific propositions, and the assumed ability of language to encompass all of the relevant subjects and predicates, the syllogistic framework should be able to be applied across all disciplines. Put another way, the expressive completeness which Aristotle intended for his theory was dependent on the epistemic primacy of the propositional form with which the theory dealt, plus a certain view of the ultimate expressive power of the range of possible terms which could substitute for his variables. These basic assumptions had to remain in place if the syllogistic was to keep its position in the intellectual hierarchy as the most general reasoning art. And, by and large, they did remain in place, and Aristotle's dogma of subject-predicate form continued as the paradigm for logic well beyond the eclipse of the ancient philosophy of science upon which it was based.

So we can see that the use of metalinguistic variables in sentence schemata, and the consequent capability of such schema to easily admit of substitutions from arbitrary subject matters, was an extremely important technique for the development of logic. Any theoretical revision to the fundamental position of categorical propositions, and hence to the corresponding sentences in which the variables were embedded, ran the risk of endangering the expressive completeness which was at the heart of logic's disciplinary identity. This fact, I believe, fostered a conservative atmosphere in which alternate notations for the syllogism were not seen as serious avenues for research.

Let us conclude this section with a final observation about how this approach to logic ended up influencing the subsequent development of the field. As mentioned above, the use of variables and schematic sentences provided the ancient logicians with a robust metalanguage for referring to the logical principles themselves. The traditional conception of this metalanguage included, in addition to the necessary (but usually implicit) rewriting directives, only variables of different types and a fixed set of logical terms. These logical terms were called the *syncategoremata* by the medieval logicians, and have come to include the familiar logical connectives "and," "or," "not," "if," "implies," and sometimes "equals;" the quantifiers "some" and "every;" and occasionally modal expressions like "possibly" or non-standard quantifiers like "most." Given this classification, then, it is easy to take the study of logic to be simply the study of admissible patterns of the *syncategoremata* within the metalanguage. This view, which Barwise has called the *first-order thesis,* is common in contemporary logic, although early versions of it can be identified in the activities of the medieval logicians and the criticisms they sustained during the Renaissance.[7]

We can observe that adopting the first-order thesis amounts to acceding to a subtle shift in Aristotle's original demarcation of the subject matter of logic. At its beginnings, logic was concerned with specifying patterns of correct reasoning that scientific reasoners might perform in the course of their activities, and so at base it revolved around actual practices. The con-

[7] See Barwise 1986, pg. 37.

ception of logic implicit in the first-order thesis, however, is that logic is primarily about specifying the properties of a particular representation which reasoners might employ, and only secondarily about providing an idealized account of the actual practice of scientific reasoning. Thus, the advantage provided by the use of linguistic variables was a two-edged sword. On the one hand, by creating a ready way to refer to inference schemata rather than individual inferences, it provided an important technical basis for logic to develop. On the other hand, though, its very power encouraged the view that the proper subject of logic was the metalanguage of the schemata themselves, rather than the actual activity of reasoning which the schemata were designed to model. To the extent that the first-order thesis was held by researchers in logic, then, metaphysical questions about the adequacy of pure subject-predicate form to capture what was common to all valid reasoning in the sciences could be relegated to an important, but unquestioned, background assumption.

9.3 Early Diagrams for Syllogistic Logic

How can we use the preceding discussion to inform our analysis of the use of diagrams in syllogistic logic? Let us turn to the historical record. From Aristotle's time until roughly the eighteenth century, there were very few types of diagrams proposed as serious alternate representations for any of the individual syllogistic forms. The diagrams which do exist fall into three broad classes: general diagrams which purport to show the relationships between terms which are characteristic of an entire syllogistic figure; derivatives of this kind of diagram which are able to represent individual moods within a figure; and diagrams which portray the immediate logical relationships between the basic types of categorical statements, such as the well-known medieval square of opposition. Because they were not intended to be used in constructing inferences, let us refer to these early types of diagrams as *static* diagrams.

The introduction of these static diagrams into syllogistic theory was undoubtedly a consequence of the clarity and ease of apprehension which their inventors saw in them. The use of diagrams made the technique of syllogistic easier to explain and summarize, and in a time of intense superstition, the possibility of representing pieces of syllogistic theory using a mysterious, regular picture must have been attractive. However, beyond taking advantage of these cognitive benefits in the presentation of the theory, no formal use was made of the distinctive graphical properties of these diagrams. That is, although derivative claims could be made about their ability to universally represent all syllogisms, these sorts of static diagrams did not themselves directly support the sorts of transformations required to model all of the varieties of reasoning in the underlying domain. Unlike Aristotle's schematic sentences, there was no system under which these diagrams could be manipulated in order to show that their represented syllogisms were valid. Hence, these sorts of static diagrams remained a teaching novelty, and outside of the mainstream of research in logic.[8]

Let us examine these early static diagrams a bit more closely. Aristotle himself appears to have employed diagrams, now lost, to illustrate the linkages between the major, minor, and middle terms of a syllogistic figure. An indication of this is in the *Prior Analytics*, where he writes:

> I call the Middle Term the one that is within another and has another within it; in the lay-out it has the middle place.[9]

In this quote, there is no sensible piece of text to which Aristotle could be referring to with the term "lay-out." We find another apparent diagrammatic reference in Aristotle's discussion of the second-figure syllogism, where he writes:

> ... what is predicated of both I call the Middle Term; what this is predicated of, the Extremes; the extreme lying nearer the middle, the Major Term; the one lying further from the middle, the Minor Term. The middle is placed outside the terms and first in position.[10]

Here, the use of the terms "position" and "placed" cannot refer to a logical relation which the middle term bears to the major and minor; hence, it is likely that it referred to some vanished diagram. Kneale and Kneale 1962 suggests, based on similar diagrams found in other ancient manuscripts, that

---

[8] A significant pseudo-logical medieval use of diagrams is contained in the 13th-century *Ars Magna* of Ramon Lull. Lull employed elaborate systems of diagrams, often with mechanical aids, in order to represent different combinations of attributes. He then applied the results to reasoning in theology and the sciences. Although Lull claimed great generality for his method, and it did involve operations performed directly on his diagrams, it is difficult to call it a logic in Aristotle's sense. For a description of Lull's method, see Gardner 1958, ch. 1.

[9] Aristotle, *Prior Analytics*, 25b36.

[10] Aristotle, *Prior Analytics*, 26b37. A very similar quote occurs at 28a12, where Aristotle discusses the third figure syllogism.

Aristotle must have been referring in these quotes to a set of diagrams for the three syllogistic figures, looking something like this:[11]

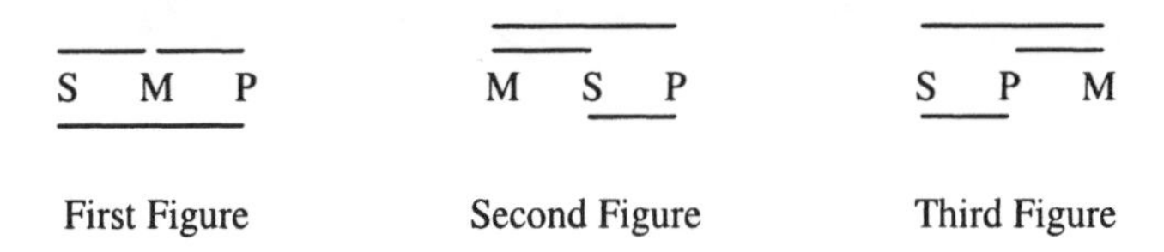

Figure 4: Aristotle's Diagrams for the Syllogism

In these diagrams, M signifies the middle term, S and P signify the major and minor terms, the upper two lines signify the term relations expressed by the premises, and the lower line signifies the term relations in the conclusion schema of the figure. We can see how M would be in the "middle place" for the first figure and "first in position" for the second figure syllogistic forms. Kneale and Kneale also speculate that if Aristotle had diagrams of this sort in mind, it would explain why Aristotle limited himself to only three syllogistic figures, because there is no other way that the middle term could be positioned in accordance with his convention that the major term must always precede the minor.[12] If true, this would comprise a sort of graphically-based completeness argument over the possible syllogistic figures. However, we have no clear evidence for this intriguing suggestion.

Whatever Aristotle's original diagrams looked like (or even if they existed at all), diagrams of this basic design were extremely popular during the Middle Ages as mnemonics and aids to explanation for the figures of the syllogism. Gardner reproduces one such diagram, due to Giordano Bruno, which attempts to combine the diagrams for each of the three figures, and superimposes them on a circle:[13]

[11] Kneale and Kneale 1962, pp. 71-2.
[12] Kneale and Kneale 1962, pp. 71-2.
[13] Gardner 1958, pg. 30.

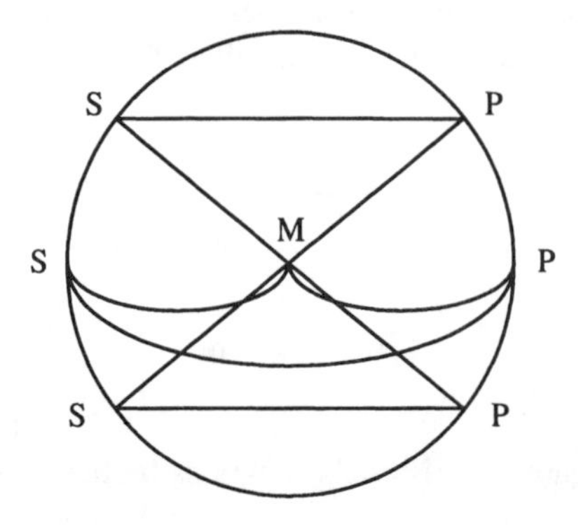

Figure 5: Bruno's Diagram for the Syllogism

This diagram apparently combines top-to-bottom position and left-to-right position to signify the various inclusion relations between the major, minor, and middle terms in the three figures, but the precise interpretation is admittedly a bit obscure. Venn also briefly discusses some more perspicuous variants of this class of diagram, in which some of the lines are given letters to signify quantification or negation, and thereby identify a particular mood instead of an entire syllogistic figure.[14] Two examples of these mood diagrams which Venn cites are:

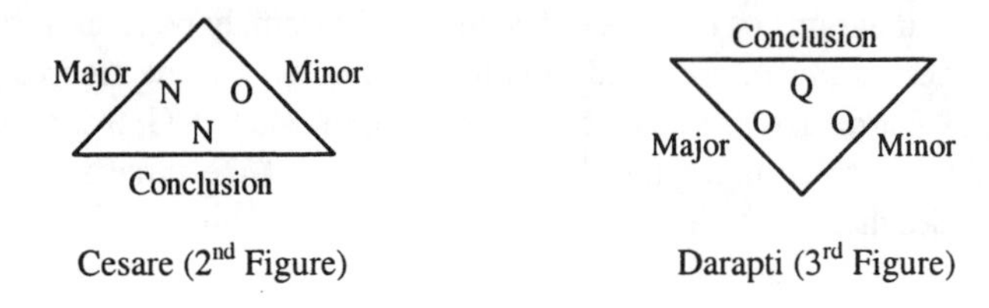

Figure 6: Two Syllogism Diagrams

The use of triangles in these diagrams may also explain their appearance in Bruno's diagram. The letters N, O, and Q in these diagrams are not vari-

[14] Venn 1884, pp. 505ff.

ables, but stand for the Latin terms *nullum* (none), *omne* (all), and *quoddam* (some), and signify the degree of quantification of the subject. Presumably, there is also a way to specify negation of a predicate term, as in the Baroco or Ferio syllogisms, but Venn does not describe it. Instead, he quickly dismisses these diagrams, saying:

> It is obvious that in diagrams of this description no kind of analysis of the proposition is attempted, and it cannot be claimed for them that they afford any real aid to the mind when dealing with trains of reasoning. For the last two or three centuries they have been entirely abandoned...[15]

Here, Venn makes two points, paralleling the two types of completeness which we earlier identified in the traditional formulation of the syllogism. First, he wonders about the match between the propositional data of logic and the representational powers inherent in these sorts of diagrams. The type of diagrammatic propositional analysis which he has in mind as sufficient is the sort given by Euler diagrams, where the proposition is understood as asserting a relation between classes of objects, and inclusion between concepts is depicted graphically as spatial inclusion. Second, Venn points out that this kind of diagram has no ability to support a reasoning system of the sort that syllogistic theory includes, and so is not a "real aid to the mind." In his view, such static diagrams are purely representational. Again, his model for this comparison is Euler's diagrammatic system, where the representation is embedded in a system which can be used to actually compute consequences, similar to traditional syllogistic theory.

The other major use of graphics in logic prior to Euler's work was the square of opposition. The first thing to notice about the square of opposition is that the deductions it represents are not syllogistic; rather, they depict certain direct inferences which can be made between two propositions only, without the interaction of a middle term. Thus, it targets immediate inference, instead of the mediate variety addressed by syllogistic theory. The background and basic distinctions in this system was given in Aristotle's *On Interpretation*, where again Aristotle apparently employed a diagram to aid his exposition in the text: "Thus there will be four cases. What is meant should be clear from the following diagram."[16]

---

[15] Venn 1884, pp. 505-6.

[16] Aristotle, *On Interpretation*, 19b28. As with Aristotle's other diagrams, the precise one referenced here has been lost, although it is fairly easy to reconstruct the square of opposition out of Aristotle's subsequent examples.

The traditional square of opposition was derived from Aristotle's discussion of the relationships between the four types of categorical sentences. It was most highly developed by the schoolmen, who ascribed elements of religion and mysticism to it, and often drew it in an extremely decorated and elaborate manner. Here is a more sparse modern rendering of the square of opposition, with the different categorical sentences annotated by their medieval letter names:[17]

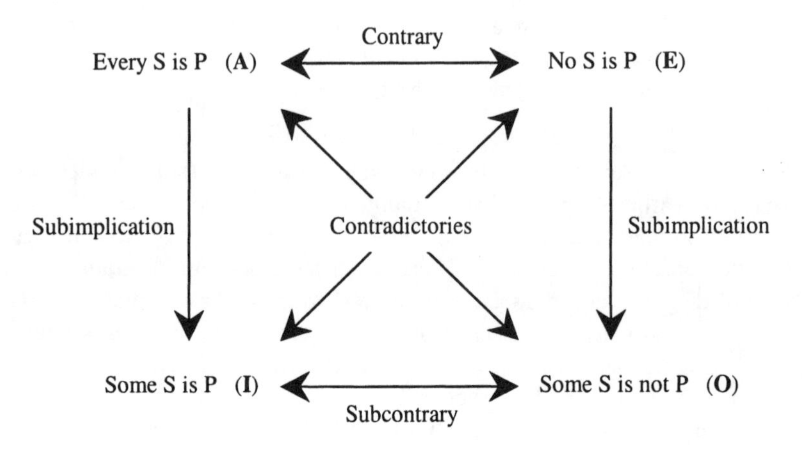

Figure 7: The Square of Opposition

The interpretation of this diagram is as follows. The four categorical sentence schema are arranged at the points of an imaginary rectangle, and the logical relationships between them are given by labeled, directed line segments. If two sentences are contradictories, then if one of the sentences is true, then the other one must be false, and vice versa. If two sentences are contraries, then they cannot both be true, although they could both be false. Two subcontrary sentences cannot both be false, but they could both be true. Finally, the relationship of subimplication is that of normal implication be-

[17] The diagram is from Salmon 1989, pg. 271. She also gives an interesting account of the origin of the letters which were used by medieval logicians to denote the different categorical sentences:

> **A**, which represents the universal affirmative, is the first vowel of the Latin "Affirmo," or "I affirm." **I**, which denotes the particular affirmative, is the second vowel of the same Latin word. **E**, which identifies the universal negative, is the first vowel of the Latin "Nego," or "I deny," and **O**, which represents the particular negative, is the second vowel of the same Latin word.

See Salmon 1989, pg. 269.

tween the top sentence and the bottom (sub) one: if the top sentence is true, then the bottom sentence cannot be false.

None of the preceding diagram types will concern us very much. The original Aristotelian figure diagrams are simply schematics which illustrate the abstract term orderings and linkage patterns which are shared by all of the schema in a syllogistic figure; they do not include facilities for either negation or for subject quantification, and so they cannot represent any individual piece of reasoning. For this reason, they do not qualify as an alternate notation for the syllogistic. Both the square of opposition and the medieval mood diagrams to which Venn referred go a little further: they can depict actual inference schemata. The sort of inference (subimplication) shown in the square of opposition, however, is so limited as to not be very useful. More importantly, however, neither type of diagram was ever seen as part of the base theory of a reasoning system. They appear to have been used primarily as static notational devices for remembering the logical relationships between the categorical propositions (for the square), or the sequence of terms and logical particles in a particular syllogistic schema (for the mood diagrams). In contrast, the sentence schemata of Aristotle's original theory were completely embedded within a larger reasoning system, upon which claims to expressive and deductive completeness could rest. They used explicit variable instantiation to directly model their subject matter in the world, and could themselves be manipulated and conjoined in order to validate more complex reasoning jumps. As such, they exhibited a larger theoretical perspective than these kinds of diagrams, and were much more tightly integrated with the activities that logic was intended to model.

9.4 Euler Diagrams and the Rise of Extensional Logic

In the eighteenth century, the great Swiss mathematician Leonhard Euler invented a completely new type of diagram: one which used spatial inclusion to represent the relationships between the subject and the attribute of a categorical proposition, and which could be manipulated in simple ways in order to show that a proposed syllogism was valid or invalid. This second aspect of Euler's diagrams marked a significant departure from the earlier diagram types, because it showed that diagrams were not limited to functioning solely as static representational devices. With these diagrams, Euler proposed an integrated notation and manipulation system which hinted at the sort of deductive completeness over propositional reasoning that was exhibited by standard syllogistic logic. This may be one reason why Euler's technique, once discovered, was so rapidly adopted. We find Venn writing that:

> Until I came to look somewhat closely into the matter I had not realized how prevalent [Euler's diagrams] had become. Thus of the first sixty logical treatises, published during the last century or so, which were consulted for this purpose ... it appeared that thirty-four appealed to the aid of diagrams, nearly all of these making use of the Eulerian Scheme.[18]

Venn also gives Euler complete credit for introducing circle diagrams into logic, writing:

> That it was practically Euler who introduced these devices into Logic, there can be no doubt: in the sense that before his time they are never to be found in the ordinary manuals, and that since that time they have been more and more frequently introduced into such treatises.[19]

Importantly, however, Euler's invention was dependent on a new way to interpret the standard categorical sentences of Aristotelian logic. During the eighteenth and nineteenth centuries, the inability to use the classical syllogistic framework to analyze certain well-known types of reasoning forms, as well as long-standing difficulties with the process of syllogistic conversion, had given some momentum to a class-relation view of the propositions addressed by the syllogism. The correct interpretation of Euler circles hinged on this view, which was ultimately dependent on an extensional interpretation of the concepts about syllogistic propositions. As we shall see, the gradual acceptance of the usefulness this extensional view of concepts in reasoning constituted a major break from the earlier Aristotelian tradition, and because of its linkages to set theory and the foundations of mathematics, was itself an important turning point in the history of logic. From our perspective, however, the major effect of using concept extensions in reasoning was to make possible an entirely new sort of diagram type for logic, characterized by the use of regions to represent concepts. But, before considering these broader implications of the change in philosophical viewpoint, we should briefly examine the specific system Euler actually proposed.

Euler described the system in a series of letters he wrote in February of 1761, although his circles did not become a common feature of logic books

---

[18] Venn 1884, pg. 110 fn. 1.

[19] Venn 1884, pg. 510.

until they were used in a successful logic treatise published in 1803.[20] Euler himself was well aware that the interpretation of his diagrams depended on a class-based view of the propositions of logic, and that coupled with this view, his circles could be reliably used as aids to deductive reasoning. Furthermore, as we will see from Euler's own words, he is careful to claim that his figures refer directly to *propositions*, and not the standard Aristotelian sentences. This tells us that Euler did not view his system as a simple graphical translation of these sentences; rather, he saw his circles as a genuine alternate notation for logic, on par with the familiar sentential forms, and not in need of these forms as a semantic intermediary in order to refer to the core categorical propositions. In this way, we can see that Euler's circle-based notation substantially narrowed the theoretical gap between the graphical notations for the syllogistic and the traditional sententially-based notation.

Euler produced separate diagrams for each of the four categorical sentences which can occur in a syllogism. In a sentence's diagram, individual labeled circles correspond to the subject and the predicate, and the area enclosed by each circle indicates the extension of the concept associated with its term. Also, presumably to prevent ambiguity in his diagrams over the proposition represented, Euler always draws the circle associated with the subject term's circle in the leftmost position. A straightforward example of his notation is Euler's diagram of the universal affirmative proposition, which he described as follows, using the traditional Aristotelian proposition that "all men are mortal" as his example:

> Thus for the notion of *man* we form a space, [reference to a circle labeled A], in which we conceive all men to be comprehended. For the notion of *mortal* we form another, [reference to a circle labeled B], in which we conceive every thing mortal to be comprehended. And when I affirm *all men are mortal*, it is the same thing with affirming that the first figure is contained in the second. Hence it follows that the representation of an affirmative universal proposition is that in which the space A, [reference to a figure like ours below left], which represents the *subject* of the proposition, is wholly contained in the space B, which is the *attribute*.[21]

---

[20] See Euler 1846, letters 102-108. Except where noted, all emphasis in quoted material appears in the original. The letters were written to the Princess of Anhalt Dessau, who was a niece of Euler's major employer and patron, the King of Prussia. The 1803 text referred to above was by the German logician K. C. F. Krause, and it is cited in Venn 1884, pp. 514-15.

[21] Euler 1846, pp. 339-40.

From this description, we can see that the Euler diagrams which correspond to the universal affirmative and universal denial propositions are:

Figure 8: Two Euler Diagrams

The particular affirmative and the particular denial are represented by partially intersecting circles, with the position of subject term label relative to the other circle signifying the containment relation between the subject and the predicate which is asserted by the represented proposition:

Figure 9: Two More Euler Diagrams

Euler's explanation for the peculiar use of letters in this diagram is brief and not very helpful:

> In affirmative particular propositions, as, *some A is B*, a part of the space A will be comprehended in the space B, [reference to a figure like above left]; as we see here, that something comprehended in the notion A is likewise in B.[22]

[22] Euler 1846, pg. 340.

The Euler diagram of a categorical syllogism is built by merging the diagrams of the three constituent propositions so that there is only a single circle per concept, but all the spatial inclusion relationships found in the individual proposition diagrams are preserved. Three labeled circles are drawn to correspond to the concepts in the major, minor, and middle positions, and the overlapping areas between them signify the degree to which the associated concepts are asserted to overlap. (Because a syllogism does not have a unique subject term, the convention of placing the subject term leftmost must be abandoned.) Here are diagrams for Barbara and Celarent, using S, M, and P for the major, middle, and minor terms:

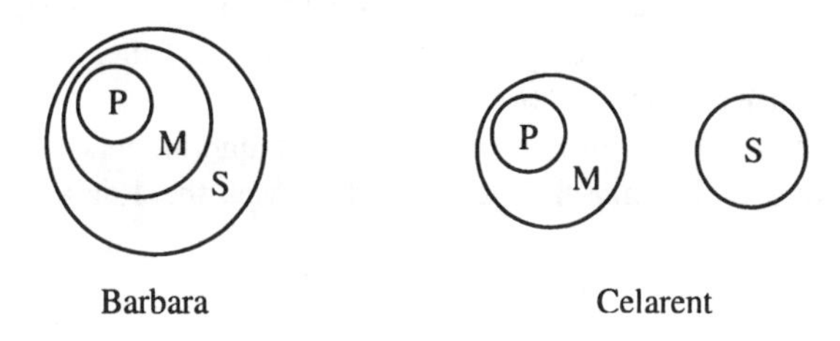

Figure 10: Euler Diagrams for Barbara and Celarent

In this way, Euler's syllogism diagrams are designed to simultaneously represent the conceptual relations which hold in each of the component propositions in the syllogism being diagrammed.

For our purposes, the further details of the operation of Euler's diagrammatic system are not crucial. Shin gives a reasonably thorough critique of its shortcomings, based on Venn's assessment and Peirce's later work.[23] For example, she notes that Euler diagrams cannot directly represent syllogisms like Disamis, because there is no way to combine the component propositional diagrams so that each concept is represented by only one circle. Euler's writings indicate that he was aware of this kind of issue in his system, but he is vague about whether he considered it a problem or not. It is also worth remembering that Euler himself was fairly modest about any explicit usability claims for his system, typically pointing only to its ability to represent syllogisms in a cognitively appealing way, as in:

> These four species of propositions may likewise be represented by figures, so as to exhibit their nature to the eye. This must be a great assistance to-

[23] Shin 1994, ch. 2.

> ward comprehending more distinctly wherein the accuracy of a chain of reasoning consists.[24]

Apparently Euler was correct in this aspect of his system, for as we have noted, Euler diagrams did become popular, and their various technical limitations eventually inspired Venn and Peirce to improve on them.

But, as we have emphasized, the important issue for us is Euler's interpretation of the propositions and concepts which he is representing with his circles, and the way in which this interpretation played into the reasoning structures he proposed. At the time Euler invented his figures, the formal distinction between the intension (comprehension) and extension of a concept was barely one hundred years old.[25] However, Euler was clearly sympathetic to an extensionally-based interpretation of propositions and concepts as a basis for asserting the validity of entailments which involved them. In one of his initial letters, he discusses the fundamental way he views the reference of general concepts:

> These signs or words [pieces of language] represent, then, general notions, each of which is applicable to an infinite number of objects: as, for example, the idea of hot, and of heat, to every individual object which is hot; and the idea or general notion of *tree* is applicable to every individual tree in a garden or a forest, whether cherries, pears, oaks, or firs, etc.[26]

Euler refers back to this view when he gives his fundamental justification for using figures to represent concepts:

---

[24] Euler 1846, pg. 339.

[25] The initial distinction between the comprehension and extension of a term was given in the famous *Port Royal Logic* of 1662 (Arnauld 1662):

> The *comprehension* of an idea is the constituent parts which make up the idea, none of which can be removed without destroying the idea. For example, the idea of a triangle is made up of the idea of having three sides, the idea of having three angles, and the idea of having angles whose sum is equal to two right angles, and so on. The *extension* of an idea is the objects to which the word expressing the idea can be applied. [pg. 51].

See also Kneale and Kneale 1962, pp. 318ff. They observe that occasional extensionally-based interpretations of terms are implicit in logical treatises at least as far back as Boethius, and perhaps earlier.

[26] Euler 1846, pg. 337.

> As a general notion contains an infinite number of individual objects, we may consider it as a space in which they all are contained.[27]

It is unclear what overall degree of acceptance this sort of extensional interpretation of concepts had among logicians at the time Euler was writing. Certainly the basic distinction between intension and extension was known. But, for logicians immersed in the philosophy of logic laid out by Aristotle, the extension of a concept would have borne very little logical weight relative to the intension. For Aristotle, it was the relations between concept *intensions* which ultimately guaranteed the validity of the syllogistic forms, and not the relations between extensions. However, by the time of Boole and Venn, it is clear that a significant, although comparatively small, group of logicians had embraced an extensional, class-based interpretation of the component concepts in a categorical proposition, and were comfortable taking the basic notion of valid inference to be defined in terms of relations between extensions, rather than the more traditional relations between intensions.

This change of philosophical viewpoint is extremely important in the history of diagrams in logic, and indeed in the history of logic generally, and so bears additional explanation. Interpreting the subject and predicate terms of a categorical sentence as referring to their extensions, and thus interpreting these sentences as expressions in a simple logic of classes, represents a significant departure from Aristotle's original stance regarding the sorts of relations in which the validity of the syllogistic was to be grounded. Let us review the basics of Aristotle's position. For Aristotle, the subject of a logical proposition always referred to some individual substance or genus of substances, and that subject could be further quantified in order to refer to part or all of the idea comprehended by it. The predicate simply indicated a not-further-analyzed attribute which was related to the subject, with the proviso that the attribute would ultimately classify into one of the categories described in the *Categories*. So, in a categorical proposition the substances referred to by the subject term were thought of as possessing a wide variety of independently-existing attributes, and the role of the proposition was simply to state the relation of one or more additional attributes to that set. Aristotle gives us some clues about his intended interpretation of predicates in a long passage from the *Topics*:

[27] Euler 1846, pg. 339.

> Next, then, we must distinguish between the categories of predication in which the four above-mentioned [genus, definition, property, and accident] are found. .... It is clear, too, on the face of it that the man who signifies what something is signifies sometimes a substance, sometimes a quality, sometimes some one of the other types of predicate. For when a man is set before him and he says that what is set there is a man or an animal, he states what it is and signifies a substance; but when a white color is set before him and he says that what is set there is white or is a color, he states what it is and signifies a quality. .... Likewise, also, in the other cases; for each of these kinds of predicate, if either it be asserted of itself, or its genus be asserted of it, signifies what something is; if, on the other hand, one kind of predicate is asserted of another kind, it does not signify what something is, but a quantity or a quality or one of the other kinds of predicate. Such, then, and so many, are the subjects on which arguments take place, and the materials with which they start.[28]

This passage is frustrating in many ways; however, we can safely conclude from it that for Aristotle, syllogistic predicates are not to be interpreted as composite, extensionally-interpreted objects. Rather, the *Categories* define a classification of the primitive types of things in the world, and among those things are substances, quantities, qualities, and so on. These sorts of objects, as Aristotle says, are the basic materials from which arguments must begin.

Aristotle's writings are certainly ambiguous about the details of his account of the reference of the terms in categorical sentences, and different aspects of his theory have been actively debated for centuries. It is not my intention here to defend an absolute position on this matter. However, it at least seems clear that for the purpose of setting out the sorts of actual relations whose interactions were to be captured by syllogisms, Aristotle could not have intended anything like the Eulerian, extensionally-based interpretation of the claims made by general statements. As we have earlier suggested, Aristotle's reason for this has to do with his fundamental reasons for claiming that syllogisms capture actual patterns of interaction between substances and their causes in the world. Take, for example, the following simple syllogism:

All swans are birds
All birds are feathered
———————
All swans are feathered

[28] Aristotle, *Topics*, 103b20-39.

As we earlier discussed in chapter 8, Aristotle holds that the reason that this syllogism expresses a correct and reliable relationship between the genus of birds, the genus of swans, and the quality of featheredness is because it is of the Barbara form, and that all syllogisms which have the Barbara "form of words" must be correct because they are derived from basic structural truths about the possible relations which primary substance can bear to an attribute. Thus, the guarantees of validity for syllogistic derivations are rooted in Aristotle's fundamental metaphysics about the world, and require that the claims made by syllogistic sentences be interpreted as claims about entities in this metaphysics. However, if these sentences are instead interpreted as claims about the inclusion relations between classes of entities (as Euler does), rather than about the predication relations between substances and their attributes, then there is no longer any Aristotelian reason to believe that the syllogistic forms continue to capture necessarily valid patterns of reasoning. Put another way, facts about the relationships which hold among such classes would not *ipso facto* be sufficient grounds to claim that the sentential relations licensed by such facts will express necessarily valid relations in a world defined by an Aristotelian metaphysics and ontology. Such a claim would require an additional argument about why truth assertions which are grounded in the inclusion relations which hold among concept extensions will necessarily track the truth assertions which are grounded in the relations between substances and their attributes. Recall that Aristotle grounded his doctrine of the syllogistic on extensive arguments in the *Categories*, *Prior Analytics*, and *Posterior Analytics* about the latter type of assertions. And, although Aristotle may have agreed that the notion of a concept's extension made sense and could occasionally be useful, he would not have agreed with Euler's position that patterns of relations between these extensions could be relied on to yield generally valid reasoning of the sort he was trying to capture with the syllogistic. For Aristotle, the fundamental reason for the validity of the syllogistic forms was *not* a result of the patterns found in the extensions of the constituent concepts, and so Euler's deductive mechanisms (which were based on such extensions) would have appeared essentially misguided.

More direct evidence for the lack of fit between Euler's system and the traditional Aristotelian conception is not hard to find. We can observe that if we interpret terms as referring to the intensions of substances or properties, then Euler's scheme for the representation of these terms by closed curves is not at all intuitive, because the possible intersections of the enclosed areas may not naturally correspond to any operation in the target domain of concepts. Take, for example, the following Celarent syllogism and its matching Euler diagram:

No unipeds are bipeds
All one-legged men are unipeds
———————————————
No one-legged men are bipeds

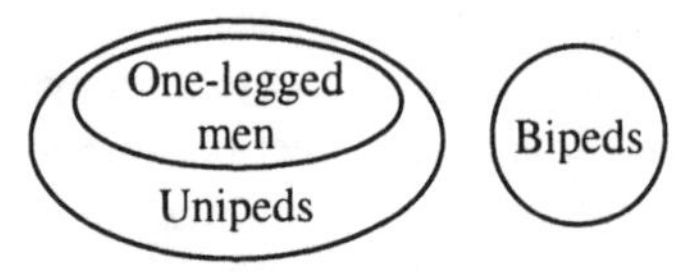

Figure 11: Comprehension of Concepts

Thinking in terms of the comprehension of concepts, how could we interpret the diagrammatic fact that the "one-legged-men" circle and the "biped" circle do not intersect? A natural thing to suggest is that we should conclude that the concepts of being a one-legged man and being a biped share no common subconcepts, because there is no area enclosed by both of the circles. Unfortunately, this is not right: in this example, both of these concepts are specializations of the underlying concept of being a creature with legs, and so involve some shared essential subconcepts (like possessing a leg), even though these subconcepts are not made explicit by the specific reasoning at hand. The basic problem here is that if a concept is viewed as indivisible, then a subregion of the circle which represents it cannot be used to signify a part of that concept; but if the concept is decomposable into a set of comprehended concepts and the area of the circle is used to signify these components, then there may be problems (like those just suggested) with the interpretation of circle intersections in a specific piece of reasoning. For these reasons, representing the intensions of concepts by the Euler's proposed system of bounded areas is likely to be awkward.[29]

The preceding example shows that Euler's system cannot be consistently interpreted as describing relations between the Aristotelian intensions of concepts. However, if we follow Euler's directions and interpret his diagrams as referring to extensionally-defined concepts in a logic of classes, this sort of difficulty vanishes. Under this interpretation, both the subject

---

[29] Indeed, if the concept A is comprehended within the concept B, then the extension of A will in general be a superset of the extension of B. As an illustration of this, consider Arnauld's example of the concept of a triangle (which comprehends the concept of having three sides): the class of three-sided objects is a proper superset of the class of triangles, because has three sides but is not a triangle since one side is not straight. This suggests that any graphical formalism based on concept comprehensions would have to be structured in a way precisely opposite from Euler's system.

and the predicate of a categorical proposition are thought of as denoting the class of objects in the appropriate concept's extension, with the sentence as a whole asserting a type of inclusion relationship between these classes. In the corresponding Euler diagram of the proposition, the two circles represent the extensions of the subject and predicate concepts, and a labeled overlap between the circles indicates an asserted overlap between the extensions. Regions are delimited by the boundaries and intersections of the circles, and are interpreted as representing the (possibly empty) class of objects which is included in the extension of the concepts represented by all enclosing circles. But, the most significant thing to note about this account of the relationship between Euler's figures and the underlying concepts is that it secures a form of expressive completeness for his system, derived from Euler's basic views about the workings of language.

Euler was impressed by the ability of language to form concepts by abstracting from groups of individuals, linking this facility to the "perfection" of a language:

> A language is always so [more perfect] in proportion as it is in a condition to express a greater number of general notions, formed by abstraction [out of sets of individuals]. It is with respect to these notions that we must estimate the perfection of a language.[30]

We can observe that the circles of Euler's system have this ability as well, because a circle can be used to represent any describable class concept. Euler explicitly makes use of this in his discussion of the possible propositions his system of circles can represent:

> We may employ, then, spaces formed at pleasure to represent *every general notion*, and mark the subject of a proposition by a space containing A, and the attribute by another which contains B. The nature of the proposition itself always imports, either that the space of A is wholly contained in the space B, or that it is partly contained in that space; or that a part, at least is out of the space B; or, finally, that the space A is wholly out of B.[31]

This quote also shows that Euler accepted Aristotle's basic theory about the four elementary forms of the categorical proposition.

---

[30] Euler 1846, pg. 338.

[31] Euler 1846, pp. 340-41, emphasis added.

Given our previous discussion of the linkage between Aristotle's theory of the syllogistic and his underlying metaphysics, however, Euler's position on the nature of the proposition in this quote seems confused. As long as categorical propositions are able to be treated as assertions about the class relations between the groups of objects referred to by the constituent concepts, Euler apparently held that his system would be able to express all of the logically relevant propositions, because his system reflected "the nature of the proposition itself." His brief explanation of this is a clear reference to the doctrine of the *Categories*, however, so it also seems here that Euler is explicitly linking the expressive power of his system to the Aristotelian system of the four categorical propositions. But, as we have observed, Aristotle's arguments for the expressive completeness of his categorical propositions are based on a particular metaphysical view about the possible relations between substances and attributes, and these arguments are not applicable to Euler's interpretation without further changes. Euler's reference to the containment of spaces demonstrate that he is talking about the relations of objects which are in the extension of a general concept, and these relations may or may not give rise to the propositions of the *Categories*. Hence, the grounds for believing that Euler's four diagrams are expressively complete are problematic.

Beyond expressive completeness, however, Euler also seems to have believed that his diagrammatic system possessed a form of deductive completeness. However, as with Aristotle's similar claim, Euler's assertion of this was largely implicit in his discussion; he had neither the vocabulary nor the conceptual framework necessary to make such an argument explicitly. We can find evidence in his writings that he held two important subclaims related to the property of deductive completeness for his system: that there can be no valid reasoning which is not, at root, syllogistic; and that his figures include all of the valid syllogisms. For Euler, the combination of these two claims would entail deductive completeness. With regard to the first, Euler went beyond Aristotle's more limited position, and suggested that reasoning via the traditional syllogistic forms would be sufficient to encompass all valid reasoning, not just scientific explanation:

> This [reasoning via the syllogistic forms] is likewise the only method of discovering unknown truths. Every truth must always be the conclusion of a syllogism, whose premises are indubitably true.[32]

[32] Euler 1846, pg. 350.

Unlike Aristotle, Euler does not attempt any sort of metaphysical argument for this claim; it seems to have been more of a reference to the philosophical dogma of the time. In view of Euler's inability to cite Aristotle's traditional arguments for this claim, because of his use of extensional interpretations of the categorical sentences, we can see the lack of an argument for this position is a critical philosophical weakness in Euler's work. In order to strengthen his position, Euler must provide reasons for believing that relations between extensions, of the sort that his deductive mechanisms model, are sufficiently powerful and reliable to yield "every truth" as the conclusion of a sequence of Euler graphs. Perhaps because of the implausibility to his student of this bare unargued claim, Euler pointed out in a later letter that the presence of syllogistic techniques in actual reasoning is frequently implicit, although he specifically asserts that syllogistic reasoning underlies formal inference in geometry:

> Hence you perceive how, from certain known truths, you attain others before unknown; and that all the reasonings by which we demonstrate so many truths in geometry may be reduced to formal syllogisms. It is not necessary, however, that our reasonings should always be proposed in the syllogistic form, provided the fundamental principles be the same. In conversation, in discourse, and in writing, we rather make a point of avoiding syllogism.[33]

We can conclude that Euler believed that all valid reasoning has its roots in the syllogistic forms, and because he based his system on these forms, he probably also believed that the deductions his system licensed would be valid. Again, though, without an argument showing how the extension-based reasoning he proposes relates to Aristotle's framework for the syllogism, Euler's position is little more than an article of faith.

The second deductive completeness subclaim which we can find in Euler's work is that his system exhausts the possible valid syllogistic forms. Euler used a combinatorial argument over the possible graphical representations for three categorical sentences and their propositional referents, and arrived at nineteen possible valid forms of syllogistic reasoning. These nineteen forms corresponded to the fourteen original schema of Aristotle, plus the five fourth-figure moods added by Aristotle's student Theophras-

---

[33] Euler 1846, pg. 354. In another letter (#108), Euler discusses reasoning with compound and hypothetical propositions. For example, he discusses the inference, "every substance is body or spirit; but A is not body; hence it is spirit," and refers to this type of reasoning as syllogistic. He makes no mention of diagrams. Because of this, Euler's use of the term "syllogism" in this quote and the previous one should probably be interpreted broadly.

tus.[34] Euler showed that each of these traditional syllogistic forms had correlates in his graphical system, although several of them could not be represented with a unique diagram. Interestingly, Euler's argument was based in part on observations about the possible ways in which his graphical representations for the categorical propositions could combine. This led him to speculate directly about the semantic underpinnings of his graphical system:

> The foundation of all these forms is reduced to two principles, respecting the nature of *containing* and *contained*: (1) Whatever is in the thing contained must likewise be in the thing containing, and (2) Whatever is out of the containing must likewise be out of the contained. Thus, in the last form [Darapti: every A is B, every A is C, thus some C is B], where the notion A is contained entirely in notion B, it is evident, that if A is contained in the notion C, or makes a part of it, that some part of notion C will certainly be contained in notion B, so that some C is B.[35]

This is the earliest place where I have found an explicit discussion of the linkage between the logical characteristics of a piece of reasoning and the topological properties of a graphical representation of it. Put in a modern phrasing, I believe that Euler is here trying to explain why the graphical features of his diagrams will correspond to the semantic relations between the concepts they represent. Euler's argument here is admittedly not very good; a more adequate discussion would address the ontological groundings for how notions can contain each other, and why reasoning based on containment relations between circles will necessarily track truth relations between the correlated concepts. And, such a discussion would lead immediately into the difficulties we discussed earlier with combining extensional interpretations of concepts with the metaphysical framework of traditional syllogistic theory.

These problems notwithstanding, Euler diagrams were an extremely important step in the evolution of diagrammatic systems of logic. Euler's was the first graphical system introduced in logic which had the potential to possess the two central characteristics we have identified as essential to the power of the traditional sentential systems: expressive completeness and deductive completeness.[36] And, although Euler's system was not satisfactory

[34] See Kneale and Kneale 1962, pp. 100ff.

[35] Euler 1846, pg. 350.

[36] The import of the critiques given by Venn and Shin is to show that Euler's original system probably was not deductively complete, even given Euler's strong Aristotelian assumptions which restricted valid deduction to that representable via the standard syllogistic forms. In particular, the inability to represent certain syllogisms by a unique Euler diagram entails

in several respects, it served as the inspiration for several more sophisticated systems, including Venn's. However, before examining Venn and Peirce's modifications to Euler's system, we should address a final important point. Recall that the main theme of this book is to provide an account for why diagrammatic systems were never part of mainstream logic. Even given the promise of Euler's work, his diagrammatic system was never accepted as part of the mainstream. It was only very recently, with the work of Shin, Hammer, and others, that Euler's diagrammatic system was seen as an independent logical system in its own right, and a serious research competitor to the traditional sentential framework. For most of its history, no body of theory was developed which used Euler diagrams in a significant way. According to Venn, the use of Euler diagrams was primarily confined to illustrations in logic textbooks. So, in view of the theoretical credentials we have just outlined for Euler's system, why was the actual use of it confined to this area?

We can suggest four possible reasons why research logicians did not adopt Euler diagrams as a tool. First of all, Euler's system would not have exhibited any compelling advantage over the established system structured around sententially-based forms for the sorts of research questions which were prevalent at the time. The scope of his system was limited to validating certain kinds of mediate reasoning between the traditionally-identified categorical propositions – the most elementary and thoroughly understood subfield of conventional logic. Euler's system did not contribute any new results to this subfield, nor did it promise to do so. His work was not designed to facilitate research on any of the more prominent problems in logic at the time, such as the analysis of hypothetical, modal, or relational reasoning. Second, the forms of actual reasoning which were taken as paradigmatic by logicians of the time remained exclusively sententially-based. Language was, and still is, the most powerful and flexible tool with which arbitrary pieces of reasoning can be expressed. A logical theory which was built directly on relationships between sentences was therefore certain to be seen by logicians as more broadly applicable. Third, although Euler's system can be helpful in illustrating the conceptual relationships which occur in a single syllogism, it rapidly becomes cumbersome and limiting when it is applied to chains of reasoning consisting of multiple syllogisms. In these cases, even its presumed advantage of cognitive clarity is lost.

Probably the most important reason why Euler's system was not more widely adopted, however, is related to the broader issues in the rise of extensional semantics, the metaphysical confusions we have alluded to at the

---

that there would exist valid chains of syllogistic reasoning for which there would be no unambiguous representation in Euler's system.

base of Euler's system, and the inertia and historical power of traditionally-sanctioned definitions in the subject matter of logic. We have previously stressed how the workings of Euler diagrams depend on an extensional interpretation of those concepts and propositions on which logic operates, and we have argued that this represented a significant change from the traditional conception of the roots of validity. For logicians working in the tradition of Aristotle, it would have been very clear that, although this new underlying domain of classes was structurally very similar to Aristotle's domain of substances and attributes, it was *not* that domain. By reinterpreting the traditional categorical sentences to refer to relations between classes of objects, extensional semantics effectively reworked the metaphysics underlying syllogistic logic, and introduced a new set of problems and issues for logic to consider.

The use of extensional semantics as a foundation for deductive systems, and all the changes in logical dogma which that entailed, provoked a very conservative reaction among logicians. In order to appreciate the depth of these changes, let us briefly consider some of the consequences which arose from this difference in subject matter. The traditional system entails that the ontology of reasoning encompass a collection of naturally-occurring substances, qualities, quantities, and the like. The subject of a proposition is distinguished from the attribute, and simple propositions assert that the subject possesses or does not possess this attribute. The relations of interest between the subject and attribute are described by concepts like "predicated of" and "inheres in," and appropriate research questions involve, *e.g.*, the appropriate conversion of predicate-terms to subject-terms and whether this entails that the predicate of a categorical proposition should be quantified.[37] On the other hand, the new class-based interpretation asserts that the ontology of reasoning consists entirely of individuals and definable groupings of them.[38] Thus, in this approach, the historic distinction between subject

[37] These two, in particular, were popular research directions in syllogistic logic during the nineteenth century. The doctrine of the quantification of the predicate involves replacing Aristotle's four categorical sentence schemata with a different set, typically five to eight in number, in which both the subject and predicate are quantified. These "quantified-predicate" schemata would include forms such as "all A is all B" or "some A is not some B." A theory of the syllogism built on a quantified-predicate framework could be seen to have certain technical advantages over the traditional formulation. The English logician Hamilton was a major nineteenth-century proponent of this sort of approach.

[38] Venn places no limitations on the flexibility of the logician to construct classes, writing: "it is of no consequence, for our purposes, whether the things which we denote by *x* and *y* are actually marked out to us by a substantive or by an adjective; by reference to their essential or accidental attributes; by a general connotative term, by a merely denotative term, or by some purely arbitrary selection of a number of individuals." Venn 1884, pg. 35.

and attribute is minimized; the proposition merely asserts that two classes of individuals bear a certain relation to one another. The relations of interest between the two classes are described by concepts like "subset" and "mapping," and appropriate research questions involve, for example, how to define hypothetical classes and limitations on the universe of discourse. To a professional logician of the eighteenth or nineteenth century, it would have been obvious that, although valid reasoning in one domain may be isomorphic to valid reasoning in the other, the two logical approaches differ substantially in their basic view of their subject matter. Studying the relations between classes simply would not have been sufficiently general or well-grounded to be seen as a part of the study of logic, even though the expression of these relations in language employed typically logical terminology. Rather, class logic would appear to be more properly a part of mathematics, especially the emerging field of set theory. For this reason, many of the traditional logicians of the eighteenth and nineteenth centuries approached deductive systems based on extensional semantics with conservatism and suspicion.[39]

---

[39] We will see some specific evidence of this in our discussion of Venn's system in chapter 10.

10

Diagrams for Symbolic Logic

10.1 Introduction

After Euler, John Venn made the next major advance in diagrammatic systems for logic. However, in order to properly evaluate the type of innovation which Venn's system represented, it will be necessary to first understand the enormous structural changes in logic brought about by the mid-nineteenth century accomplishments of George Boole. Section 10.2 will therefore briefly review that portion of Boole's work which was relevant to Venn's diagrammatic system. Boole's reputation in logic rests on his 1854 book, *An Introduction to the Laws of Thought*, in which he provided the first detailed example of the use of a formal, mathematically-inspired deductive method in logic. This book stimulated the development of symbolic techniques in logic, and is generally considered to be the first work in symbolic logic. More importantly for our purposes, however, the generality of Boole's approach to logic offered a new way to understand logic's proper scope and subject matter, and this had a direct effect on the development of diagrammatic representational systems for logic. Boole's approach expanded the domain of logical theory to include reasoning based on the unions and intersections of arbitrary numbers of classes, and, because of this, gave the discipline of logic its first serious alternative to the doctrine of the syllogism. Venn was among the first logicians to recognize the potential of Boole's theories, and the diagrammatic system he produced was intended to be interpreted within the context of Boole's work.

Section 10.3 uses the discussion of Boole's work of section 10.2 as a background against which to evaluate the diagrammatic system described in Venn's 1881 book, *Symbolic Logic*. In this book, Venn attempted to popularize Boole's work by diminishing its overtly mathematical character, adding some simplifications of his own, eliminating its quantitative and probabilistic portions, and (unlike Boole) explicitly tying his development throughout the book to the concerns of traditional logic. We will first discuss Venn's attitude toward logic as a discipline, which was inspired by Boole's stance and shares many of the same goals. Next, using this as a background, we will explain how Venn intended his diagrammatic system to be a tool with which to illustrate a certain subclass of the sort of reasoning treated by Boolean logic, and only derivatively as a tool with which to illustrate the syllogistic. In the course of this, we will examine Venn's system in detail and contrast it with Euler's. Interestingly, because Venn was operating with a wider conception of the scope and subject matter of logic than was Euler, the criteria for evaluating the expressive and deductive completeness of Venn's system will necessarily vary from that used for Euler diagrams.

Section 10.4 will take up Peirce's system of existential graphs. Unfortunately, Peirce's contemporaries largely ignored his innovative diagrammatic logical system, and the distinctive philosophy of logic upon which it was based had no discernable impact on subsequent developments in symbolic logic. We will start by examining the early, Boolean-inspired work of De Morgan on the logic of relations, and show how this inspired Peirce's initial work on this topic. Next, we will use this context to discuss Peirce's examination of Venn's diagrammatic system, and the modifications he carried out on it. We will also see how Peirce's inability to modify Venn diagrams so that they could represent relations encouraged him to create his own diagrammatic system: the system of existential graphs. Peirce had a complex set of philosophical motivations for his choices in constructing the system of existential graphs, and we will consider them at some length. Finally, we will conclude this section by suggesting some reasons for why existential graphs, and the characteristic Peircean philosophy which accompanies them, had such a minimal effect on the other logicians of the time.

Our investigation of the role of diagrams in symbolic logic will conclude with section 10.5, in which we will make some observations on the late nineteenth-century effects on diagrammatic systems that were brought about by the development of symbolic logic in Europe. We will concentrate on the work of Frege and Hilbert. Like Venn and Peirce, these logicians were inspired by Boole's initial work in symbolic logic, but together they went far beyond Venn's relatively minor modifications to Boole's system. For our purposes, the changes in logic that they wrought can be grouped into two main categories, and these changes together spelled the end of the use of diagrams in mainstream logic. First, along with Peirce, Frege introduced relations and quantifiers into formal symbolic logic. This additional expressive power rendered Venn diagrams (as well as Peirce's modified Venn system) useless for all but the most elementary reasoning in this new logic, because the original Boolean view of symbolic logic in which Venn diagrams were grounded did not include these features. Second, as we have noted in our previous investigation of geometry, both Frege and Hilbert espoused a strong antipsychologism about logic, and indeed about rigorous proof in general. Both of these men held that our ability to interpret and use diagrams was necessarily dependent on facts about our particular facility to intuit spatial objects, and so neither one would accept that diagrams could play any formal role in a logical proof. Further, their stature in logic was such that this antipsychologistic attitude with respect to operations and representations used in logic became widely known and adopted. Section 10.5 will conclude by suggesting that the overall result of these shifts in the foundations of logic was that diagrammatic techniques were more or less ignored by the logicians following Hilbert, and that this prejudice has continued more or less unchallenged until the present day.

10.2 Boole's Symbolic Logic

Boole's highly original accomplishments in logic were based on explicit analogies between the operations of algebra and those of logic. He transformed collections of sentences into systems of logical "equations" describing the class relations in the sentences, and gave quasi-mathematical methods for operating on these equations. Boole's methods blended algebraic equation-solving with traditional logical concerns, where the elimination of class terms in the logical equations paralleled the elimination of concepts in syllogistic reasoning. However, Boole's logic was able to analyze far more sophisticated forms of reasoning than the classical syllogistic, and as a result he had very ambitious goals for his system. He explicitly hoped that his work would become the foundation of a new science of logic that would incorporate the familiar syllogistic patterns, but would ultimately extend over far broader and more general types of reasoning. To this end, Boole designed his system to encompass both categorical and hypothetical reasoning involving complex propositions, and additionally to validate reasoning involving probability relations. Concerning the relation of his work to the Aristotelian syllogistic, he wrote:

> The course which I design to pursue is to show how these processes of Syllogism and Conversion may be conducted in the most general manner

> upon the principles of the present treatise, and, viewing them thus in relation to a system of Logic, the foundations of which, it is conceived, have been laid in the ultimate laws of thought, to seek to determine their true place and essential character.[1]

This quote illustrates two important themes in Boole's work. First, it is clear that Boole intended to pursue a reductionist goal with respect to traditional logic: he wanted to show that the doctrines of traditional logic result from the operation of the more general laws identified by his system. So, according to Boole, the patterns of validity identified by the syllogistic are a consequence of other, more basic logical laws which his work makes explicit. Second, this quote shows that Boole was explicitly interested in deriving his logical theory from a preexisting set of internal "laws of thought." This idea runs counter to the basic metaphysics underlying Aristotle's approach to logic: that logical truths are grounded in an external metaphysical theory of the possible relations between a group of preexisting individuals and properties. This approach, that the primary task of a logical theory is to describe the fundamental preconditions to coherent thinking, had been implicit in logic for most of its history, but became much more powerful in the eighteenth and nineteenth centuries after the controversies surrounding the work of Hume and Kant on the core notions of metaphysics. And, given the metaphysical problems surrounding Euler's diagrams in logic, it is important to note that Boole's appeal to the laws of thought offers at least the hope of supplying class logic with a suitable foundation.[2] However, we should note that Boole's appeal to the laws of thought as a base for his logical theory is a bit disingenuous. Although Boole explicitly linked his logic with the "constitution of the intellect" in the final chapter of his book, the actual development of his theory proceeded without any significant reference to mental processes, and Boole's followers quickly recognized that his system was in fact based on the realization that the pure calculus of algebra could be generalized so as to structure non-numeric relationships. Thus, by using the framework of Boole's system, it became possible for the study of logic to be separated from psychology, theories of mental faculties, or indeed to any features of the epistemic situation of the reasoner.

---

[1] Boole 1854, pg. 228.

[2] As an important logician in the nineteenth century (Boole held the chair of mathematics at Queen's College, Cork), it is safe to assume that Boole would have been familiar with Euler's class-based formulation of the syllogistic. However, Boole never mentions Euler's work as an influence on his own. Rather, his work in logic seems to have been inspired by his contemporaries and more direct predecessors, such as De Morgan and Hamilton.

This separation caused Boole's conception of logic to be strikingly modern, and prepared the way for Frege's attempt to remove human intuition from the processes of mathematical proof. Boole envisioned his basic logical system as a sparse and highly abstract language of class symbols and operators, augmented by a set of formal rules whose operation was not dependent on the specific interpretation of the class symbols. These rules described the possible class expressions and give methods for transforming them and eliminating symbols from them. With a system structured in this way, Boole believed that he could put logic and mathematics on the same philosophical footing, and avoid many of the metaphysical and linguistic entanglements that had been characteristic of previous work in logic. More interestingly, though, he can be seen to be nevertheless following out the consequences of Aristotle's dictum that the scope of logic was to be defined by what was common to all reasoning, and so that the structures and methods of logic themselves cannot contain any elements which would tie them to a particular subject matter.

The import of Boole's work was to extend Aristotle's vision of the scope of logic along two main dimensions. First of all, Boole's theory expanded the form of the propositions to which logical theory would be directly applicable. In his theory, the traditionally central role of the standard subject-attribute categorical proposition was replaced by the broader notion of a "primary proposition:" one which could be interpreted as asserting a combinatorial relation determined by the unions and intersections of two or more classes of individuals. This generalization alone represented a major step in the evolution of extensional logic, replacing the previous simple binary logic of classes with more sophisticated mechanisms to address these combinatorial functions of arbitrary numbers of classes. Additionally, Boole's system defined a class of secondary propositions: those which assert relations between primary propositions. He was able to use this two-tiered structure to define valid reasoning with hypothetical, disjunctive, and probabilistic propositions, and consequently was able to consolidate into one system many of the separate areas of nineteenth-century logic. As we shall see, because of the new propositional forms treated by his system, the expressive ability of Boolean notation went significantly beyond what was representable with Euler diagrams, because Euler diagrams were based on representing only simple categorical propositions involving two classes.

The second way Boole's system extended Aristotle's vision of logic was that he provided unified methods for reasoning with all of his targeted types of propositions. These techniques, which were inspired by algebraic procedures for solving systems of equations, were far more powerful than the various different reasoning schemes which had been proposed for the categorical, disjunctive, hypothetical, and probabilistic propositions. As we have observed, though, Boole's methods achieved their power while remaining independent of the interpretation of the individual class symbols, and so his system conformed to Aristotle's requirement that logic not include reasoning methods which were not generally applicable to any subject matter. Also, the regularity of these methods suggested to several nineteenth-century logicians that Boole's system could be used as the foundation of a mechanizable general calculus of logic. The quest for a universal technique for the automatic computation of consequences goes back at least to the thirteenth-century Lullian Art which we mentioned earlier, although its most systematic development prior to Boole was probably in Leibniz's late seventeenth-century attempt at formulating an *ars combinatoria* containing a *calculus ratiocinator*.[3] The application of Boole's methods, however, led the British logician William Jevons in 1869 to produce an important forerunner to modern computers: a machine which mechanically solved simple Boolean logic equations.[4]

A brief example will both give a flavor of Boole's method and make the foregoing discussion clearer. We will outline one of Boole's simpler examples, taken from his analysis of Samuel Clarke's "Demonstration of the Being and Attributes of God." Clarke was a lesser-known contemporary of Spinoza, and structured his philosophical writings in a way similar to those of Spinoza, including explicitly identifying his premises and conclusions. Because of this characteristic, Boole found both Clarke's and Spinoza's writings to be sources of useful illustrations for his method. One specific proposition of Clarke's which Boole analyzed is that, "the unchangeable and independent Being must be self-existent." Clarke claimed that this conclusion follows from the following three premises, which he argues for elsewhere in his writings:

1. Every being must either have come into existence out of nothing, or it must have been produced by some external cause, or it must be self-existent.
2. No being has come into existence out of nothing.
3. The unchangeable and independent Being has not been produced by an external cause.

[3] Interestingly, Leibniz apparently believed that the *characteristica universalis*, the language which would underlie the *ars combinatoria*, would be iconic, rather than based in a purely sentential notation. See Kneale and Kneale 1962, pg. 328.

[4] Jevons' logic machine, and Marquand's 1881 improvement of it, are described in Gardner 1958, pp. 91ff.

Boole was able to demonstrate that, in his logical system, Clarke's desired conclusion follows from these premises. For the purposes of our example, we will here provide only an outline of Boole's proof, and skip many of the details surrounding his exact rules of inference and techniques for representing propositions.[5]

Boole's notation revolves around symbols for not-further-analyzed classes. He uses "1" to denote the class which contains all individuals, and "0" to denote the class which contains no individuals. For any class symbol s, its complement is denoted $(1 - s)$, its intersection with another class t is denoted st, and its union with t is denoted $s + t$. Thus, in Boole's system, we have laws like $x(1 - x) = 0$ and $0x = 0$. Deduction proceeds via a set of rewrite rules which are explicitly designed to be similar to the ones employed in standard equational reasoning in algebra. For example, from the expression $x(1 - x) = 0$, Boole can derive (in a way parallel to standard equational reasoning) that $x - xx = 0$, and so that $x = xx$, which Boole believes to be a fundamental law of thought called the "law of duality."[6] However, Boole's system also includes several rewrite rules which would not be valid when applied to traditional mathematical equations.

As a first step in rendering Clarke's argument in his system, Boole defines a set of class terms corresponding to the classes which arise from Clarke's original formulation:

x = Beings which have arisen out of nothing.
y = Beings which have been produced by an external cause.
z = Beings which are self-existent.
w = The unchangeable and independent Being.

Using these variables, Boole symbolizes Clarke's three premises (listed above) in the following way:

1. $x(1 - y)(1 - z) + y(1 - x)(1 - z) + z(1 - x)(1 - y) = 1$
2. $x = 0$
3. $w = v(1 - y)$

Some complexity is already apparent here with Boole's formulation of premise (3), in which he employs his special "indefinite class" symbol v in order to represent the class of objects which is indefinite in all respects except that some of its members have not been produced by an external cause.

[5] Further details of this particular proof can be found in Boole 1854, pp. 194-95ff.

[6] The expression $x = xx$ is interpreted that the intersection of any class with itself is equal to the original class. See Boole 1854, pp. 49-51.

The effect of the v in (3) is to indicate that w is not equal to the entire class defined by $(1 - y)$, but only a portion of it, and therefore that (3) makes a particular claim. As the first step in reasoning with these representations, however, Boole must eliminate the indefinite term in (3), and so Boole uses a selection of his rules governing indefinite classes in order to rewrite (3) into:

4. $wy = 0$

Next, by using (2) to eliminate the occurrences of x in (1), Boole arrives at:

5. $y(1 - z) + z(1 - y) = 1$

By expanding (5) and taking the complement of each side, Boole gets:

6. $yz + (1 - y)(1 - z) = 0$

By combining (4) with (6) and eliminating the variable y, Boole gets:

7. $w(1 - z) = 0$

In order to make his conclusion match the Boolean form of Clarke's original conclusion, Boole reintroduces the indefinite variable v. By doing this, he is able to rewrite (7) as $w = vz$, which is Boole's translation of Clarke's original conclusion.

10.3 Boolean Logic and Venn Diagrams

As we can see from the foregoing, one result of Boole's decision to use the calculus of algebra as his model for the calculus of logic was that he freely employed algebraic notation to signify class relations. This approach proved to be disastrous for the broad acceptance of Boole's work. For most classically trained logicians, the major difficulty with Boole's work was that its highly mathematical character (for the time) made it completely unrecognizable as logic, and consequently further exacerbated their overall skepticism toward extensional interpretations of concepts.[7] Venn, however, was

[7] Venn also suggests that, "much of the prejudice which for some time existed against the employment of symbolic methods in Logic must be attributed, unless I am mistaken, to the extreme length and elaboration with which Boole generally worked out his results." Venn 1884, pg. 347.

an exception to this group. His 1881 *Symbolic Logic* was an extended attempt to provide a less technically demanding introduction to the non-probabilistic segment of Boole's logic. He also simplified Boole's system by introducing some refinements of his own, and placed the whole treatment in a framework which was substantially closer to the concerns of traditional logic than was Boole's own work. In his effort to support Boole's approach, Venn devoted several chapters of his book to an extended discussion about the ways in which general operations could be abstracted away from their origins in a particular class of operands, and specifically defended Boole's use of algebraic operations in a logical context. Venn's explicit target with this work are those logicians who are prejudiced against, as he puts it, "any work professing to be a work of Logic, in which free use is made of the symbols + and -, × and ÷."[8] As an example of this attitude, he quotes his contemporary Spaulding as saying:

> All attempts to incorporate into the universal theory of Thought a special and systematic development of relations of number and quantity must be protested against .... No cumbrous scheme of exponential notation is needed, and none is sufficient, for the actual guidance of thought when its objects are not mathematical.[9]

We can see from this quote that Spaulding's objection to Boole's work is grounded in a classically Aristotelian conception of the proper scope of logic. Spaulding believed that the techniques of mathematics would not be sufficiently general to be included in a universal science of reasoning, which would also be applicable "when its objects are not mathematical."

In contrast to logicians like Spaulding, Venn maintained with Boole that certain mathematical operations were more universally applicable than their origins in algebra would initially indicate. For example, Venn emphasized that some of the basic class-combinatory operations used by Boole to manipulate logical equations could apply to extensionally interpreted concepts generally, and so should be included in a general science of reasoning. Venn addressed Spaulding's worry about a "cumbrous scheme of exponential notation" by suggesting that:

> We may regard Symbolic logic and Mathematics as being branches of one language of symbols, which possess some, though very few, laws of com-

[8] Venn 1884, pg. ix.

[9] W. Spaulding, *Introduction to Logical Science* (1857), pg. 50, quoted in Venn 1884, pg. ix.

> bination in common. This community of legislation or usage, so far as it exists, is our main justification for adopting one recognized system of symbols for both alike.[10]

Concerning the ability to generalize the domain of application of this common system of symbols and combinatory principles, Venn had this to say:

> [T]hose who call in the aid of symbols generally find that they have got possession of a machine which is capable of doing a great deal more work, and even work of very different kinds from what they originally expected from it. To the mathematician, of course, this is perfectly familiar.[11]

However, according to Venn, the larger community of logicians gave no credence to the possibility of using in logic techniques and notation derived from mathematics, and this objection was the root of their rebuff of Boole's work. These logicians believed, with Spaulding, that the forms of reasoning used in mathematics would not be subject-neutral enough for a generally-applicable logic, and they therefore found Boole's reductionist program to be metaphysically suspect. Venn also disagreed with Boole about the scope of his reductionist program, at least with regard to the practical teaching of logic:

> No one can feel more strongly than I do the merits of [traditional logic] as an educational discipline. And this conviction is even enhanced by the fact that some of the most instructive portions of the common system are just those which Symbolic Logic finds it necessary to pass by almost without notice. Amongst these may be placed the distinction between Denotation and Connotation, the doctrine of Definition, and the rules for the Conversion and Opposition of propositions .... Common Logic should in fact be no more regarded as superseded by the generalizations of the Symbolic System than is Euclid by those of Analytical Geometry.[12]

Venn's evident concern with logic education, present in this quote and throughout his book, may have provided the initial inspiration for his dia-

---

[10] Venn 1884, pg. xvi.

[11] Venn 1884, pg. 431.

[12] Venn 1884, pp. xxvi-xxvii. In view of the history of geometry which we have described in Part I, Venn's comparison between traditional logic and Euclidean geometry, on the one hand, and symbolic logic and analytic geometry, on the other, is provocative.

grammatic system.[13] Certainly, Venn diagrams provided a simple graphical way by which traditional logicians could understand the complicated class relations described by Boole's logic. Let us now turn to a discussion of Venn's system.[14]

We will begin by making explicit an important point which has been implicit in the foregoing discussion: just as Euler's diagrammatic system was inspired by a change in the perception of the subject matter of logic, so too was Venn's. Euler's system was made possible by the introduction into logic of extensionally-interpreted concepts. We have earlier described how Euler explicitly constructed his system in order to model syllogistic relations between the four categorical propositions. The fundamental Euler diagrams were designed to represent the class relations occurring in these four propositions, and the diagram for an individual syllogism was produced by combining the diagrams of the three constituent propositions. Venn's system, because it was based on Boole's expansion of the types of propositions treated by logical theory, had to depart from a representational strategy based on the four simple categorical propositions. In order to accommodate the more powerful class structures required by Boolean logic, Venn defined a new fundamental type of diagram, called a *primary* diagram, with which to depict the class relations in Boolean sentences. Primary diagrams are composite geometrical figures which contain just enough distinct regions to represent a "definite number of classes or compartments which are mutually exclusive and jointly exhaustive."[15] Thus, a primary diagram for n classes will have 2^n regions, where each region corresponds to a different combination of inclusion and exclusion relations between the n base classes. Primary diagrams are reminiscent of Lullian diagrams in that they only display possible class combinations; they cannot themselves represent propositions or specific relations between these classes.[16] Further, different numbers of classes will require different primary diagrams to represent them. Venn gave the following example of the simplest primary diagram in his system, and contrasted it with Euler's similar diagram:

---

[13] See, *e.g.*, Venn 1884, pg. 128, where Venn discusses the usefulness of his diagrammatic system, and claims that, "it is that sort of visual aid which is [his diagram's] especial function." As we have noted, Euler made a similar claim for his system.

[14] Prior to the publication of *Symbolic Logic*, Venn had introduced his diagrammatic system in "On the Diagrammatic and Mechanical Representation of Propositions and Reasonings," in *Philosophical Magazine*, July 1880. See Venn 1884.

[15] Venn 1884, pg. 111.

[16] See Section 9.3, fn. 8, for references to the Lullian Art.

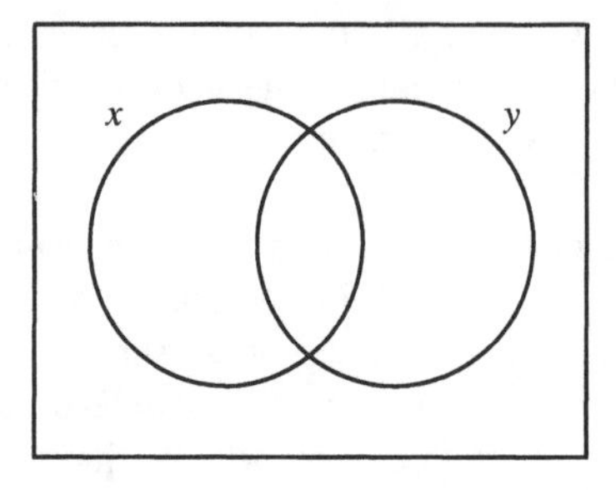

Figure 12: A Simple Primary Diagram

> On [the Eulerian scheme] this would represent a proposition, and is indeed commonly regarded as standing for the proposition 'some x is y'; though ... it equally involves in addition the two independent propositions 'some x is not y' and 'some y is not x', if we want to express all that it undertakes to tell us. With us, however, it does not as yet represent a proposition at all, but only the framework into which propositions may be fitted; that is, it indicates only the four combinations represented by the letter compounds, xy, $\neg xy$, $x\neg y$, $\neg x\neg y$.[17]

In this way, *contra* Euler diagrams, the pattern of intersections of the figures in a primary diagram does not indicate any particular class relation, and so it does not represent a proposition. Rather, as Venn suggests, it only provides the "framework into which propositions may be fitted." Other primary diagrams include the standard three-circle arrangement, as well as other less-familiar primary diagrams for representing the combinations of more than three classes. Venn's general procedure for constructing primary diagrams is:

> We should merely have to begin by drawing any closed figure, and then proceed to draw others in succession subject to the one condition that each is to intersect once, and once only, all the existing subdivisions produced by those which had gone before.[18]

---

[17] Venn 1884, pg. 114. Technically, Venn is incorrect in calling this an Euler diagram, because Euler always placed the class symbol within its corresponding circle.

[18] Venn 1884, pg. 118.

With this scheme, each figure that is drawn represents a class, and the graphical regions that are thereby produced represent the complex class defined by the presence of all the enclosing classes and the absence of all the non-enclosing classes. We must remember to include in our regions the one enclosed by none of the figures. It has been recently shown that primary Venn diagrams can be constructed using this basic technique for any number of classes.[19]

The function of the primary diagram in Venn's system also demonstrates to us that Venn, like Euler, intended his diagrammatic system to directly represent class relations in a target domain, and not just to serve as a graphical notation for a set of logically prior Boolean equations. The appropriate primary diagram in a Venn representation is determined solely by a semantic fact: the number of logically relevant classes of individuals that are present in the context of reasoning. Thus, the determination of the primary diagram is made in much the same way as the selection of the proper number of class symbols in a Boolean modeling framework, and so is independent of the choices made in any specific Boolean interpretation. In explaining his diagrams, Venn rarely appealed to preexisting Boolean formulations of propositions; rather, he spoke generally about ways to determine the proper primary diagram from the problem at hand. Because of this, we can see that Venn most likely envisioned his diagrams to be on a symbolic par with a certain subclass of Boolean equations, and a genuine alternate notation for that part of the logic.

Venn's system represents universal propositions by shading different subregions of a core primary diagram. A shaded subregion corresponds to an assertion that the complex class associated with that subregion is empty. To guarantee that all universal propositions would result in a pattern of shading, he argued that every universal proposition is equivalent to the denial of one or more existential claims, and that the denial of an existential claim corresponds to the emptiness of a particular class of individuals defined by the claim.[20] So consider, for example, a universal categorical proposition describing the relation which all elements of a given class bear to the elements of some other class or classes. In order to represent this claim, Venn would first draw a primary diagram which could represent all possible combinations of the classes referred to in the proposition. Then, he would convert the universal proposition into an equivalent particular claim asserting that certain class combinations were empty, and shade those re-

[19] See Polythress and Sun 1972.

[20] See Venn 1884, pg. 123. This basic relationship between universal and existential propositions has been known since Aristotle, and is symbolized in modern logic by the quantifier equivalence $(\forall x\ \varphi) \leftrightarrow (\neg\exists x\ \neg\varphi)$.

gions in the primary diagram. For example, to represent the quantified-predicate proposition 'all x is all y', Venn would draw the following two-class, four-region primary diagram:[21]

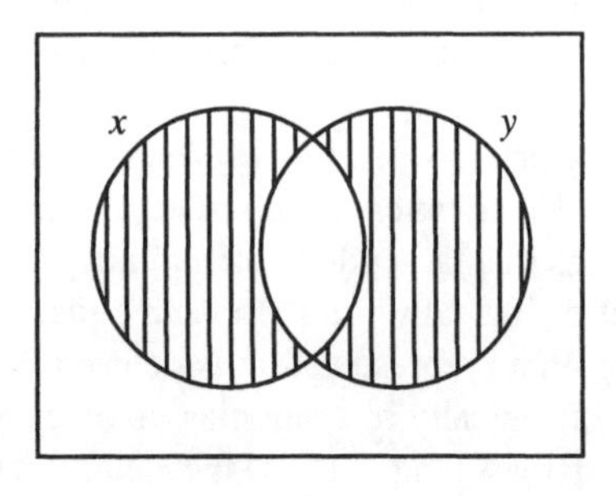

Figure 13: Primary Diagram for "All x is all y."

Venn constructs this diagram by converting 'all x is all y' to the two propositions, 'all x is y' and 'all y is x'. The first of these is interpreted as claiming that there are no individuals in the class of things which are both x and not-y, and so that region is shaded. The second claim is interpreted in a similar way, and the corresponding region in the primary diagram is shaded. We end up with the diagram above, in which two regions are shaded, $x\neg y$ and $\neg xy$, and two regions are not shaded, xy and $\neg x\neg y$.

Like Euler, Venn also wanted to represent valid pieces of reasoning with his diagrammatic notation. In order to do this, he employed the property of his system that it can use the same primary diagram to represent multiple propositions which each assert the emptiness of a different class combination. We saw an illustration of this in our previous example, where a single two-class primary diagram was used to represent both 'all x is y' and 'all y is x'. Venn's strategy for representing reasoning involving only the sort of propositions described above was to successively represent each of the premises on a single primary diagram, and then extract the conclusion from the resulting diagram of the combined premises. Venn gives us an example of Celarent, one of the traditional first-figure syllogistic forms which employs only simple universal affirmative and negative propositions:

---

[21] See Venn 1884, pg. 122. We have previously noted (ch. 9 fn. 37) that the doctrine of the quantification of the predicate, from which Venn's example proposition was drawn, was an active research area in logic at the time Venn was writing.

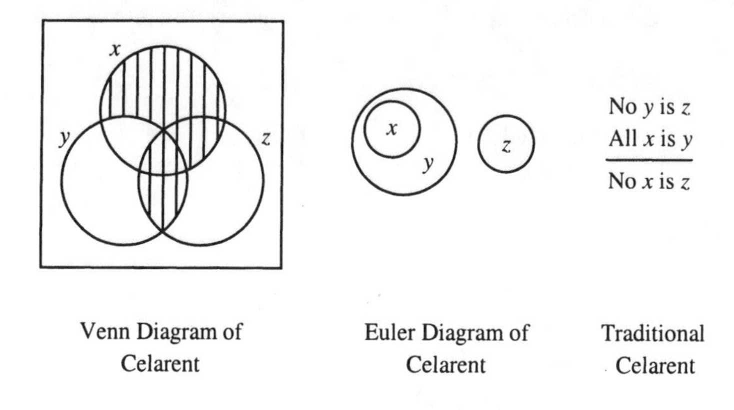

Venn Diagram of Celarent

Euler Diagram of Celarent

Traditional Celarent

Figure 14: Three Diagrams for Celarent

In this example, Venn admitted that it was not clear that his diagram was more perspicuous than Euler's.

Venn's major point in his book was that both Boole's system and his diagrammatic representation were designed to handle far more sophisticated types reasoning than the traditional syllogistic. Therefore, in spite of his support for the use of the syllogism as an educational tool, Venn was largely dismissive of the practical utility of traditional syllogistic reasoning. Given the tools of common logic, he rhetorically asks whether or not, "we ever fail to get at a conclusion, when we have the data perfectly clearly before us, not from prejudice or oversight but from sheer inability to see our way through a train of logical reasoning?"[22] Furthermore, Venn saw clearly that much of the classical dogma of the syllogism was inapplicable to the new sort of symbolic system which Boole had created. Venn wrote compellingly about the differences between the two approaches:

> We must frankly remark that we do not care for this venerable structure [of the syllogistic] .... its ways of thinking are not ours, and it obeys rules to which we own [sic] no allegiance. To it the distinction between subject and predicate is essential, to us this is about as important as the difference between two ends of a ruler which one may hold either way at will. To it the position of the middle term is consequently worth founding a distinction on, to us this is as insignificant as is the order in which one adds up

[22] Venn 1884, pg. xx.

> the figures in an addition sum. On the other hand the distinction between universal and particular propositions which to it is unimportant is to us vital.[23]

Because of this attitude, and his overall goal of clarifying and popularizing Boole's work, the reasoning examples which Venn supplies are generally more powerful than can be analyzed via the classical syllogistic. An example is the following:[24]

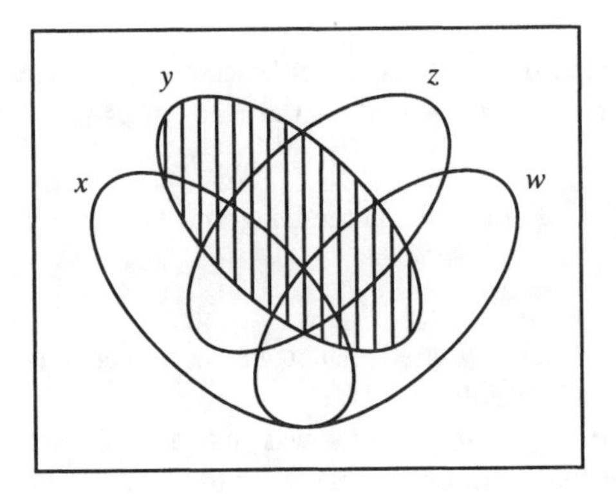

Figure 15: A Four-Class Venn Diagram

What obvious conclusion follows from:

1. Every *y* is either *x* and not *z*, or *z* and not *x*.
2. Every *wy* is either both *x* and *z*, or neither of the two.
3. All *xy* is either *w* or *z*, and all *yz* is either *x* or *w*.

Venn's diagram (Figure 15) plainly shows the desired conclusion, that there are no individuals in the class *y*.

Venn's book contains an entire chapter of detailed examples of Boole's method, in which many of them are solved via diagrams. None of his examples are directly solvable via the syllogistic. We should also note that Venn's own diagrammatic system is not the only one in use in Venn's book; Venn also discusses and uses Marquand diagrams, which he concedes are more useful than his system when the number of classes exceeds four or

---

[23] Venn 1884, pp. 402-3.

[24] This example is from Venn 1884, pg. 129.

five. Marquand diagrams are essentially tables of class-combinations, and as far as I know had no further impact on logic.[25]

However, even given the representational and reasoning power which Venn diagrams exhibited relative to Euler diagrams, Venn diagrams suffered from major technical problems which significantly limited their influence on the logical practice of the time. As with our discussion of Euler diagrams, we will structure our evaluation of these problems around the twin notions of expressive completeness and deductive completeness. First, Venn's system was severely deficient in expressive power. Most obviously, as Venn admitted, his system was extremely awkward relative to the standard sentential notation when representing reasoning involving more than five classes and their complements. Although it is technically possible to construct the requisite primary diagrams for reasoning with six or more classes, the use of such diagrams is exceedingly cumbersome, and their graphical complexity causes them to frequently be at a cognitive disadvantage compared with Boole's sentential notation. Venn was aware of this issue, and promptly conceded that the visual advantages of his system break down when the number of classes exceeds five. In this situation, he speculated that some other diagrammatic system, such as the abovementioned Marquand diagrams, might be a suitable alternative to Boole's notation.

Beyond this cognitive issue, however, is the structural problem that Venn's system has no way of representing that a class is not empty. This means that arbitrary existential claims cannot be represented, and therefore his system clearly is not expressively complete. This deficiency is not present in the underlying Boolean conception of logically admissible propositions; it was solely an issue with Venn's graphical system. Venn admitted that the inability to represent existential claims was a potential problem with his diagrams:

> It will very likely be objected that we are here taking account only of universal propositions, and that any system of diagrams must be prepared to represent particulars also.[26]

He briefly speculated about a possible modification to his system which might allow him to represent more general existential claims, but he did not describe this solution in detail. Rather, he apparently hoped that, in actual applications of his diagrammatic system, reasoning involving these sorts of claims would be rare. For example, he writes:

[25] See, *e.g.*, Venn 1884, pp. 137-140.
[26] Venn 1884, pg. 130.

> We must not be misled by the position which [particular claims] occupy in ordinary Logic; where the very narrow schedule of admissible propositions, and the requirements of the syllogism, seem to give them a relative frequency and importance far beyond what they really deserve. In what may be called real life, and still more in science, they take a much more humble place. Science aims always at the universal and the definite; and the true logical particular is seldom there regarded as anything more than a temporary correction of some previously accepted or proposed universal...[27]

This quote might be construed as an argument by Venn that a logical system does not have to support representation or reasoning with existential claims, and therefore that the lack of such support in his system is not a real problem. Boole also seems to have held a variant of this position:

> The logician affirms, that it is impossible to deduce any conclusion [in the physical sciences] from particular premises. .... [T]he principle of order or analogy upon which reasoning is conducted must either be stated or apprehended as a general truth, to give validity to the final conclusion. In this form, at least, the necessity of general propositions as the basis of inference is confirmed...[28]

Nevertheless, there are two obvious reasons why this would be a curious position for Venn to take. First of all, Venn clearly believed that logic as a discipline was nearly useless for the sort of reasoning which occurs in science or everyday life generally, referring to the utility of logic as "speculative rather than practical."[29] Therefore, the demands of actual reasoning would be an odd basis on which to justify a restriction of his diagrammatic system. Second, despite Boole's effort at linking logic back to scientific reasoning, much of the import of his work was to show that the truths of logic can be developed from a small number of axioms and a general symbolic method of inference, independent from any particular interpretation of the class symbols. From this (admittedly more modern) point of view, Venn's appeal to the perceived structure of actual scientific reasoning is a step toward a more Aristotelian view of logic, and away from the symbolic theories upon which his diagrams were explicitly based. At any rate, it ap-

---

[27] Venn 1884, pg. 131.
[28] Boole 1854, pp. 403-4.
[29] Venn 1884, pg. xx.

pears that Venn did not devote very much effort to giving his diagrammatic system the ability to represent particular claims.

Unfortunately, the expressive deficiencies with Venn's system are not limited to its inability to represent existential claims. The propositions which Boole admitted into his logical system also include the whole category of secondary propositions, such as the secondary disjunctive "all a is b or all a is c," and its cousin the hypothetical proposition. And, because Venn's diagrammatic system was designed to depict the class relations describable in Boole's logic, one would expect that Venn diagrams would include a technique for representing these sorts of propositions as well. Venn does sketch a modification to his system which might allow it to represent certain limited sorts of disjunctive secondary propositions, but his discussion is very tentative, and he later blames the inherent complexity of the relation itself for his troubles:

> Of course this [representing general disjunctive propositions] is complicated, but this complication lies in the nature of things, rather than the depraved ingenuity of the logician.[30]

As for the other kinds of secondary propositions, such as general hypotheticals, Venn simply gives up. Interestingly, he finds it necessary to emphasize that the ability of his system to capture even simple hypotheticals is dependent on a Boolean account of the interpretation of propositions, and that this account does not respect the traditional view of hypotheticals as referring to cognitive judgments:

> Of course when we consider the hypothetical form as an optional rendering which only differs verbally from the categorical, we may regard our diagrams as representing either form indifferently. But this course, which I regard as the sound one, belongs essentially to the modern or class view of the import of propositions. Those who adopt the judgment interpretation [of hypotheticals] can hardly in consistency come to any other conclusion than that hypotheticals are distinct from categoricals, and do not as such admit of diagrammatic representation.[31]

These significant difficulties with expressive completeness ensured that the impact of pure Venn diagrams on logical practice would be comparable to

---

[30] Venn 1884, pg. 133.
[31] Venn 1884, pg. 521.

that of Euler diagrams: they would be primarily used as illustrative devices for a limited class of problems in logic texts.

A second major issue which would have affected the role of Venn diagrams in the logical practice of the late nineteenth and early twentieth centuries was that the reasoning methods which Venn described for his diagrammatic system were only minimally developed. Venn's basic deductive technique was derived from the property of his diagrammatic system that it could only represent claims which assert that a class has no members. Therefore, since a primary Venn diagram can simultaneously represent multiple claims of this type, reasoning using Venn diagrams proceeds by successively adding to the diagram information about empty classes, and examining the final diagram in order to draw conclusions about the overall class relationships that are entailed. It is clear, though, that this reasoning method will neither capture all valid reasoning nor even all valid Boolean reasoning. In fact, as a result of the inability of Venn diagrams to represent either the particular affirmative or the particular negative categorical proposition, the pure Venn system cannot even represent the first-figure syllogistic moods of Darii or Ferio. Therefore, although Venn diagrams can be used to validate types of reasoning which Euler's cannot, Venn's system is obviously not deductively complete in the sense that the syllogistic is, or that Euler diagrams might be.

At this point, we should recall our previous observation that Boole's accomplishments in symbolic logic gave an important boost to a fundamentally different view of the philosophical linkage between logic and the world: that (*contra* Aristotle) the metaphysical basis of logic could be founded on the laws of thought, rather than in the fundamental structure of properties and relations in the world. Thus, the core Aristotelian notion of deductive completeness which was appropriate for us to apply to Euler diagrams – that a deductively complete logic will be able to capture all valid general reasoning in the sciences – is no longer suitable for us to use as a standard to evaluate the deductive completeness of Venn's system. The accepted goals and purposes for a logic in 1881, at the time when Venn was writing the first edition of his book, were quite different from those operative at Euler's time, and so our standards for evaluating Venn's theory need to change accordingly. Boole's writings in 1854 about his underlying purpose in creating his new class-based symbolic logic illustrate this transition. Boole did attempt to link his logic back to strictly scientific reasoning in the classical Aristotelian way, but mostly he emphasized that his work was intended as a broad-ranging investigation of the character of coherent thought itself. And, in the years between 1854 and the 1890s, this second metaphysical approach had gathered a small following. Consider, for example, the quote with which Venn begins the second (1894) edition of his *Symbolic Logic*:

> At the time when the first edition of this work was composed it would scarcely be too much to say that the conception of a Symbolic Logic was either novel or repugnant to every professional logician. An amount of explanation and justification was therefore called for which does not seem to be quite so necessary now.[32]

Certainly Frege and Hilbert employed the symbolic approach. Recall that Frege's 1879 *Begriffsschrift*, with its attendant philosophy of logic, was published a few years before Venn first devised his system, and Venn's citations to it show that he knew of Frege's work and respected it.[33] Because of this, when we consider whether Venn's diagrammatic system is deductively complete, we should evaluate if it would be suitable to model the consequences of the hypothesized laws of coherent thought, rather than whether it can capture all valid general reasoning in the sciences. And, although this standard as presented is admittedly too vague to be very useful, because of the essential weaknesses already observed in Venn's system, it is clear that Venn diagrams cannot meet it.

In Venn's defense, he never argues that his diagrammatic system is anything but a cognitively appealing visual aid for illustrating a particular species of Boolean reasoning. Certainly, Venn readily employs Boole's more powerful symbolic deductive methods whenever his examples demand it, and even occasionally uses Marquand diagrams to illustrate a particularly obscure set of class relations. Because of this, it is clear Venn did not intend his system to compete directly with symbolic notation in all cases. However, when the paucity of deductive power in Venn's system is combined with its lack of expressive completeness, the result is undeniably a weak system. This realization undoubtedly caused most other logicians to see Venn diagrams primarily as a notational variant for a small part of Boolean logic, and not as an independent logical system in its own right. Indeed, from the perspective of logic at the turn of the century, Venn's system must have inhabited an uneasy middle ground. To a classical logician who was already skeptical of the work of Boole, Venn's entire scheme must have looked as if it was based on a metaphysical misunderstanding. However, to

---

[32] Venn 1884, pg. vii.

[33] See Frege 1879. Note, however, that Venn would almost certainly have viewed Frege's logicist program as only one portion of the spectrum of coherent thought for which logic had to account.

a symbolic logician, it would have been clear that Venn diagrams had only a fraction of the power of Boole's algebraic representation, and thus they would not have been worthwhile to study or use.

Both of these basic problems with Venn's system – that it cannot adequately represent many types of non-universal Boolean claims, and that it includes only a meager deductive system – were noted and largely remedied by Charles Peirce in two papers in 1896, two years after the publication of the second edition of Venn's book.[34] In fact, the graphical system which is known today to most students as "Venn diagrams" is Peirce's update to the pure Venn system, and this Peirce/Venn formulation was used as the basis for Shin's more modern Venn-I and Venn-II graphical deductive systems.[35] Peirce's changes to Venn's system can be divided into two major categories. First, Peirce modified Venn's original diagrammatic representation scheme so that it was able to represent all of Boole's primary propositions. He replaced Venn's simple device of shading with a technique of annotating subregions with an **O** or an **X** to indicate information that a class was known to be empty or non-empty, respectively. He also proposed that a connecting line be drawn between these markings when it is necessary to indicate that the markings are to be interpreted disjunctively, and he described a larger graphical framework which allowed sets of two or more diagrams be able to be interpreted as disjoined with each other. With these two features, Peirce was able to claim that his new system was expressively complete relative to the power of the universe of Boolean propositions. Therefore, given Venn's assumption that the propositions encompassed by Boole's system would include all those that might be potentially relevant to a logical theory, the Peirce/Venn system appeared to be expressively complete *simpliciter*.

The second way that Peirce updated Venn's system involved a substantial expansion and rigorization of the basic deductive method Venn had proposed. Peirce eliminated Venn's original informal technique of sequentially adding class information to a primary diagram, and replaced it with six, explicit, graphically-based rules of transformation operating over the figures of the new Peirce/Venn system. These rules allowed the new system to directly validate many different sorts of reasoning, including all of the traditional syllogistic figures and moods, as well as a much more substantial fragment of the deductions licensed by Boolean logic. However, even given the increase in the inferential power of Venn's system that was imparted by the six transformation rules, Peirce apparently came to suspect that the resultant system would not be able to derive individual diagrams representing all of the possible Boolean consequences of arbitrary

34 See Peirce 1931, 4.356ff.

35 See Shin 1994.

all of the possible Boolean consequences of arbitrary premises.[36] (As Shin has demonstrated, Peirce was correct in this suspicion; however, at the time the conceptual apparatus necessary to prove this was still thirty-five years in the future.) Further, as we shall find in the next section, Peirce did not agree with the critical assumption that the expressive and deductive power of Boole's original system would be sufficient to incorporate the entire subject matter of logic. Hence, after suggesting his improvements to Venn's diagrammatic scheme, Peirce abandoned this line of thought in order to work on three newer and more radical diagrammatic deductive systems for logic which was not circumscribed by the limits Boole's work: his alpha, beta, and gamma systems of existential graphs, which we will address in the next section.

We will not describe the Peirce/Venn system in further detail. Recall that our project in this book is to trace the evolution of diagrammatic systems in logic, and explain why they had no effect on the mainstream practice of logic. And, at least at first, Peirce's modifications to Venn's system affected no one but Peirce himself. As was alluded to above, the primary reason for this was the prior realization by Peirce, as well as several of the leading logicians of the 1890s, that the basic framework of Boolean logic itself needed to be overhauled and expanded. It is ironic that Venn's system, which is an important historical event in the development of diagrammatic systems for logic, was conceived in order to illustrate a conception of logic which, even by Venn's time, was losing its currency. This fundamental Boolean view of logic, which was inherent in Venn diagrams – that the central task of logic was to use certain purely combinatory relations of classes and their complements to represent universally valid patterns of thought – was being displaced by Peirce, Hilbert, and Frege even as Venn's *Symbolic Logic* was published. So, just as the advent of the extensional view of concepts caused older diagrammatic schemes for the syllogism to give way to Euler's circles, the advent of Hilbert's mathematical logic and the first-order logic of quantifiers and relations in turn rendered obsolete both Venn's system and Peirce's update to it.

To understand this why this was so, let us again refer to parallels with the history of Euler circles. The invention of Euler circles depended on the then-new semantic idea of interpreting concepts extensionally. The power of this idea forced a change in the underlying ontology of traditional logic, and made inadequate the earlier graphical representations of reasoning, like Bruno's diagram and the square of opposition. In a similar way, Boole's invention of symbolic logic had caused the notion of a logically analyzable

---

36 See Shin 1994, pg. 28 fn. 32.

piece of reasoning to be extended beyond the traditional patterns of categorical sentences with which Euler dealt, and thereby made it possible for more complicated class relations to be included within the scope of logical theory. This change in the underlying context of logic had in turn made Euler's diagrams obsolete, because their interpretation was tied to the traditional syllogistic framework. Boolean logics of multiple classes involved arbitrarily large combinations of distinct concepts and their complements, and so required representational systems which could unambiguously denote the relationships between them. The goal of Venn's system was to do for Boolean logic what Euler did for extensional logic: to provide a visually-based system for understanding the propositional relationships entailed by the new logical approach. However, the foundations of a new and more powerful view of logic than Boole's began to emerge in the 1870s, and by 1900 the logical systems inspired by this view had overwhelmed both Boole's original system and Venn's reworking of it. Evaluating the possibility of diagrams in this new symbolic logic will be the task of our two final sections.

10.4 Peirce's Existential Graphs

One of the first people to read and understand the implications of Boole's work, even before Venn, was Augustus De Morgan. Boole had corresponded several times with De Morgan about the possibility of applying algebraic techniques to logic, and De Morgan used this fundamental idea in an influential paper of 1864.[37] De Morgan's overt goal with this paper was to provide a more rigorous, Boolean-inspired analysis of the basic logic of a certain obscure class of complex binary relations with which he was concerned at the time.[38] Arguments involving binary (and occasionally higher-degree) relations had been a subject of sporadic inquiry since antiquity, but no truly general theory had ever emerged. At best, medieval logicians such as Ockham had developed elaborate sets of methods to convert arguments involving certain common relations into arguments which contained only

[37] See De Morgan 1864.

[38] The class of relations which De Morgan studied were called "relatives," and were typically composed of the conjunction of two simpler relations. An example of one of the relatives De Morgan studied is "lover of a master of." De Morgan was interested in these sorts of relations because he thought that the correct application of a relative would be dependent on the existence of a suitable mediate term. Because of this feature, he believed that an understanding of relatives could lead to a better understanding the basic relation of predication in the syllogistic. See, *e.g.*, Kneale and Kneale 1962, pg. 428.

standard combinations of Aristotelian attributes. Using this strategy, the medieval logicians attempted to show how these sorts of arguments could be brought into the familiar syllogistic framework, and be made coherent with Aristotle's underlying ontology of substances and attributes.

De Morgan's paper turned this traditional approach on its head. Instead of bending relational arguments to fit the constraints of syllogistic form, he argued that the doctrine of the syllogistic itself was a consequence of more general logical laws which governed the valid composition of relations. Specifically, De Morgan hypothesized that, since all syllogisms were logically dependent on the workings of the basic relation of predication (*e.g.*, that a particular attribute is or is not predicable of a subject), the rules of syllogistic reasoning should be derivable from the transitive property of the simple binary relation "is," which appears in every categorical sentence. His paper attempted to provide a Boolean-styled analysis of this possibility, and discussed some of the more general "modes of combination" exhibited by different kinds of transitive relations. De Morgan's theory was ambitious but not philosophically satisfactory, because he did not explicitly address the linkages between his theoretical principles and the hypothesized structure of the world which would ultimately justify formal arguments involving relations. However, even the moderate success De Morgan achieved by using Boolean techniques to systematize and propose a calculus for the long-confusing logic of relatives was sufficient to inspire several other researchers to take an algebraic approach to this problem.

Two of the most important of these researchers were Frege and Peirce. It is one of the interesting coincidences in the history of logic that both Frege and Peirce, although working apart and having substantially different goals in mind for the logical systems they developed, arrived roughly simultaneously at many of the same innovations in symbolic logic. Motivated by De Morgan's insights, they each independently conceived of the core modern logical notions of relation function, quantifier, and bound variable. In doing this, both of them contributed to a fundamental shift in the background context in which research in logic had been conceived, away from Boole's emphasis on simple class combinations and algebraic analogies, and toward a more general, abstract view emphasizing arbitrary relations between objects. This shift had a profound effect on the possibility of diagrams in logic. In terms of overall historical impact, however, Peirce was the least influential of the two, in spite of the fact that he developed a much more original and sophisticated philosophy of logic in which diagrammatic deductive systems played an important role. In this section, we will concentrate on the accomplishments of Peirce, and proceed to Frege in our final section.

Around the time De Morgan published his paper, Peirce became familiar with Boole's work, and in 1870 Peirce wrote a paper in which he reviewed Boole's logical system and De Morgan's application of it to the logic of relatives. In this paper, Peirce was clearly disturbed by the aesthetics of the Boolean-inspired system which De Morgan had constructed:

> [De Morgan's] system leaves much to be desired. Moreover Boole's logical algebra has such singular beauty, so far as it goes, that it is interesting to inquire whether it cannot be extended over the whole realm of formal logic, instead of being restricted to that simplest and least useful part of the subject, the logic of absolute terms, which, when he wrote, was the only formal logic known.[39]

More interesting than Peirce's opinion of De Morgan's work, however, is his characterization of the changes in logic which had happened over the 15 years following the 1854 publication of Boole's book. This quote shows that during this period Peirce believed that the subject matter over which an adequate formal theory of logic must range had widened considerably, presumably to include the study of relations. This belief was certainly not universal among the logicians of the time. But, as a result of Peirce's acceptance of this change of scope, he was inspired to follow De Morgan's initial idea and use a Boolean algebraic approach as the basis for a new logic of relations which would also encompass the traditional syllogistic. Peirce generalized and improved on De Morgan's work in a series of papers which he published between 1870 and 1885, and successively refined his logical theory of relations into a system which was expressively identical to the first-order notational system in use today.

In his development of the general logic of relations, Peirce introduced two important innovations that were critical to the early success of symbolic logic. First, the way that Peirce structured his theory of relations was an important step in advancing logic beyond concerns rooted in the classical Aristotelian metaphysics of the syllogism and Boole's ontology of classes. Instead of reexamining complex questions concerning the intensional or extensional nature of classical subjects and predicates, Peirce's theory of relations accepted Boole's framework of reasoning with extensions, and instead addressed itself (as De Morgan had) to generalizing the fundamental notion of predication contained in the syllogistic. This idea was not new; at roughly the same time as Aristotle, the Stoics had also developed a theory which attempted to analyze binary relations other than simple predication,

[39] Peirce 1931, 3.45-6.

and (as we have noted) this problem appeared sporadically in the writings of the medieval logicians. However, as a consequence of the more flexible Boolean-inspired ontology in which he was working, Peirce was far more successful than any of his predecessors at systematizing the general logic of relations. Second, Peirce's work on relations led him directly to a major technical advance in symbolic logic: the introduction of explicit quantifiers and the notion of a bound variable. Because Peirce was following the mathematical intuitions of Boole and De Morgan, from the beginning he attempted to structure the calculus of his theory of relations so that it would conform to a set of certain explicitly stated quasi-algebraic rules. For example, Peirce initially conceived of his relations on analogy with arithmetic sums and products, and included analogs of certain algebraic principles, such as the laws of distribution and inverse. However, in order to achieve adequate generality for his theory, he developed notation for arbitrary relational signs and the sums and products of them, and by 1883 he had coined the term "quantifier" for two operators which he used to range over the individuals in his relative sums and products. These operators were designed to capture the logical intuitions behind the terms "some" and "every," and were essentially identical to the quantifiers used in modern logic. In fact, Kneale and Kneale 1962 characterizes the final system Peirce proposed in 1885, which he referred to as the General Algebra of Logic, as "adequate to the whole of logic and identical in syntax with the systems now in use."

This background allows us to more fully understand the context of Peirce's 1896 critique of Venn's diagrammatic system, which we discussed in the previous section. Even though his citations show that Venn was aware of the work of Peirce, Frege, and De Morgan, he did not himself address any of these new developments in symbolic logic in his book. Explicit quantifiers of the sort invented by Peirce and Frege are never mentioned, and the only reference to the logic of relations is a brief mention in the context of a discussion of Lambert's eighteenth-century work in logic. Several explanations for this omission are possible. The simplest one is to suggest that, unlike Peirce, Venn did not believe that the scope of the reasoning which could be analyzed by formal symbolic logic had been substantially broadened since Boole's time. Perhaps Venn did not believe that quantifiers, relations, and the other new devices in symbolic logic were mature or stable enough to be put in his book, which was primarily designed as a popularization of the pure Boolean theory. Another possible reason for Venn's omission is that the inclusion of a robust theory of relations would make Venn's diagrammatic system obviously inadequate to the expressive task of logic, and in order to avoid this consequence he purposely chose to limit himself to Boolean logic. However, this reason would elevate the diagrammatic component of Venn's overall treatment to a degree of importance which it is doubtful it deserves. It is quite clear that the algebraically-based logical system described in Venn's book is intended as the primary theory, and that Venn viewed his diagrammatic representation as merely a more perspicuous secondary notation, and as a convenient alternative to Boole's often dense logical equations. At any rate, Venn's decision to ignore the latest work in symbolic logic certainly made it an attractive basis for Peirce's later modifications.

From his other philosophical work, about which we will say more below, Peirce was persuaded that the existing sententially-based notations for symbolic logic were too opaque to be satisfactory for the sorts of intellectual discovery which he felt were the ultimate goal of any reasoning system. Peirce had been initially trained as a chemist and was familiar with the molecular diagrams used in organic chemistry, and he envisioned an analogous diagrammatic notation for logic which would similarly expose the logical structure of propositions. When he took up Venn diagrams in 1896, his previous work in philosophy allowed him to recognize that Venn diagrams might provide a framework for the sort of system he was looking for. Peirce's basic requirements for a diagrammatic logical system were fairly simple, and yet they show that he had a much more modern conception of the role of proof in a such a system than did Venn:

> Our purpose, then, is to study the workings of necessary inference. What we want, in order to do this, is a method of representing diagrammatically any possible set of premises, this diagram to be such that we can observe the transformation of these premises into the conclusion by a series of steps, each of the utmost possible simplicity.[40]

Peirce soon realized that, in order to achieve the goals he had in mind for a diagrammatically-based logical system, the expressive power and deductive capability of pure Venn diagrams would need to be strengthened in ways that would go far beyond the strict Boolean calculus upon which it was based. In one of his papers on the subject, Peirce identified four distinct areas of concern with Venn's existing formulation:[41]

1. Venn diagrams cannot represent existential statements.
2. Venn diagrams cannot represent general disjunctive statements.

---

[40] Peirce 1931, 4.429.
[41] See Peirce 1931, 4.356ff.

3. Venn diagrams cannot represent numbers, statistical facts, or measurements.
4. Venn diagrams cannot represent relations.

Peirce was able to address his first and the second concerns by modifying Venn's system in the way we outlined in the previous section. Peirce's third concern involved deductive capabilities which were present in Boole's original logical work (and, to a more limited extent, in Peirce's previous research), but which Venn deliberately ignored in order to make *Symbolic Logic* more accessible. As it turned out, this issue did not worry Peirce very much. However, in view of Peirce's philosophical commitments and his involvement in the initial development of the logic of relations, he saw as critical the inability of Venn's system to be modified to meet his fourth concern. Because he could see no way to amend Venn's scheme so that it could represent arbitrary relations, he completely abandoned his rework of Venn's system, and over the course of the next ten years he worked on developing his own diagrammatic system of existential graphs.

Peirce viewed his system of existential graphs as his most significant contribution to logic, and he eventually produced three separate systems: the alpha system, which has the same expressive and deductive power as modern propositional logic; the beta system, which extended alpha to include the structures of predicate logic; and the gamma system, which augmented beta with additional notation and rules to allow it to express relations in modal logic.[42] Ironically, Peirce's alpha system has been the most well studied of the three, even though it is the least powerful and is even more expressively limited than the Boolean fragment which Venn employed in *Symbolic Logic*. (We have observed earlier that Peirce viewed Boolean logic as "the simplest and least useful part of the subject.") Unlike beta and gamma, the syntactic

---

[42] Peirce describes the alpha and beta systems of existential graphs in the following way (Peirce 1931, 4.510-11):

> The alpha part of graphs ... is able to represent no reasonings except those which turn upon the logical relations of general terms. ..... The beta part ... is able to handle with facility and dispatch reasoning of a very intricate kind, and propositions which ordinary language can only express by means of long and confusing circumlocutions. A person who has learned to think in beta graphs has ideas of the utmost clearness and precision which it is practically impossible to communicate to the mind of a person who has not that advantage. Its reasonings generally turn upon the properties of the relations of individual objects to one another.

The description and motivation for the gamma system is quite complex, and involves several concepts which are part of Peirce's general philosophy of logic. In particular, Peirce seems to have tried to build a possible-worlds mechanism into the diagrammatic syntax of gamma, although he never completed it. See, *e.g.*, Peirce 1931, 4.512ff. Unfortunately, this manuscript cannot further explore this fascinating system.

flavor of the alpha system is fairly easy to convey. A set of propositional letters is indicated on a surface (which Peirce called the "sheet of assertion"), and non-intersecting closed curves surround subsets of the letters, as in:

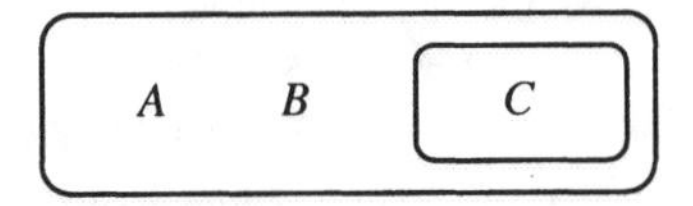

Figure 16: A Peirce Diagram for $((A \wedge B) \rightarrow C)$

The curves are interpreted as negating the objects inside them, which are themselves interpreted conjunctively. So, the graph above represents a proposition which would be written in traditional notation as $\neg(A \wedge B \wedge \neg C)$, or $((A \wedge B) \rightarrow C)$. Peirce also defined a set of graphical rules of inference for alpha and beta, similar in spirit to the rules he defined for the Peirce/Venn system. Hammer has recently shown that the syntax and proof theory which Peirce gave for alpha is sound and complete relative to a traditionally-constructed semantics.[43]

On the strength of Hammer's results, it appears that Peirce's alpha system of existential graphs is the first purely diagrammatic logical system ever constructed that we can say is both expressively complete and deductively complete relative to its *probable* intended semantics.[44] Admittedly, we cannot know for certain whether Hammer's choice of semantics for alpha captures all and only the reasoning tasks which Peirce had in mind for alpha. We do know that alpha is neither expressively nor deductively complete

[43] Hammer 1995 gives a formal analysis of the semantics and proof theory for alpha, plus citations to previous work on alpha and beta.

[44] At one point, Peirce suggested that the semantics for his logical systems was intended to at least encompass all of mathematical reasoning, although he hedged about whether or not his system would be sufficient for this purpose.

> Now a thorough understanding of mathematical reasoning must be a long stride towards enabling us to find a method of reasoning about [topology] as well, very likely, as about other subjects that are not even recognized to be mathematical. This, then, is the purpose for which my logical algebras were designed but which, in my opinion, they do not sufficiently fulfill. The present system of existential graphs is far more perfect in that respect, and has already taught me much about mathematical reasoning. Whether or not it will explain all mathematical inferences is not yet known.

See Peirce 1931, 4.428-9.

relative to the standards we applied in our earlier investigations of the logical systems of Euler or Venn, because its context, scope, and intended subject matter are quite different from those of any of its predecessor diagrammatic systems. Because of its lack of quantifiers, alpha cannot even represent the basic Aristotelian categorical sentences. Peirce's tripartite division of the system of existential graphs into the alpha, beta, and gamma fragments, together with the quite different expressive powers and deductive rules assigned to each, shows us that he had a distinctly modern hierarchical understanding of the scope and power of a logical system. For alpha in particular, it is clear that Peirce had extremely restricted goals in mind. Recall that Aristotle had hoped with the syllogism to systematize all scientific reasoning, and that Venn designed his diagrams to capture a substantial fragment of Boolean reasoning. Alpha was explicitly designed to be limited to reasoning about the basic logical relations between unanalyzed general terms, and therefore would clearly fail at the outset to capture much of the interesting reasoning in the world – reasoning which Peirce himself agreed was part of the purview of formal logic. Beta was designed to capture more of the sorts of reasoning in which Peirce was interested, and gamma may well have been complete relative to the classical subject matter. However, Peirce's description of the exact sort of reasoning he wanted to capture with these three systems was not precise. The metaphysics underlying his logical system was not specified tightly enough to determine if the alpha, beta, or gamma systems of existential graphs were either expressively or deductively complete relative to Peirce's intentions.

Perhaps the fairest thing to say about Peirce's existential graphs is that they have had only a minimal effect on the practice of logic. In fact, the various systems of existential graphs are probably even less well known than Peirce/Venn diagrams, which we recall is the system that Peirce rejected as too weak to be useful as a notation for logic. The primary intellectual impact of existential graphs has occurred in computer science, where a small group of researchers have used existential graphs as a foundation for a specialized representational system called conceptual graphs.[45] So, in view of the important technical features which Peirce's existential graphs introduced into diagrammatic logical systems – including a rigorously-specified syntax and an explicit set of deduction rules whose applicability was determined solely by the formal properties of the representation – why has their impact been so slight? We can distinguish two main sets of reasons for this: one group concerned with the technical construction of his system, and the other more philosophical.

[45] The seminal work on the use of conceptual graphs in computer science is Sowa 1984.

Let us first examine the more technical aspects of his system which would have made it unpopular with other logicians. Unlike his predecessors, Peirce explicitly disavowed many of the claims of usability and perspicuity which have motivated all of the previous graphical systems we have discussed. He acknowledged that his particular notation was not constructed with ease of teaching or cognitive accessibility in mind:

> [A]lthough the study of [existential graphs] and practice with it will be highly useful in helping to train the mind to accurate thinking, still that consideration has not had any influence in determining the characters of the signs employed; and an exposition of it, which should have that aim, ought to be based upon psychological researches of which it is impossible here to take account.[46]

Sowa has nevertheless claimed that, at least for the alpha fragment, "Peirce's graphs are very readable: with a little practice, the reader can immediately see the various combinations of boxes as disjunctions and implications."[47] So, Peirce's worry about cognitive problems may only be a weak explanation for why existential graphs were not more well-studied. However, Peirce also points out that:

> [T]his system is not intended to serve as a universal language for mathematicians or other reasoners, like that of Peano.[48]

Given the state of logic in the late nineteenth and early twentieth centuries, this is a much more serious issue. Much of the leading-edge logical work in Europe, such as that of Frege, Russell, and Hilbert, was concerned with systematizing the sorts of reasoning problems which occur in mathematics, and Peirce's deliberate choice not to tailor his system for expressing propositions in this domain or any other domain must have immediately made it problematic as a tool for real logical research.

Finally, Peirce also admitted that beyond these expressive issues, the detail-oriented nature of his system would made it *more* difficult to use as a deductive calculus than a traditional system:

---

[46] Peirce 1931, 4.424.

[47] Sowa 1984, pp. 139-40. Of course, given the nature of his project, Sowa can hardly claim anything different.

[48] Peirce 1931, 4.424.

> [T]his system is not intended as a calculus, or apparatus by which conclusions can be reached and problems solved with greater facility than by more familiar systems of expression. Although some writers (Schröder) have studied the logical algebras invented by me with that end apparently in view, in my own opinion their structure, as well as that of the present system, is quite antagonistic to much utility of that sort. The principal desideratum in a calculus is that it should be able to pass with security at one bound over a series of difficult inferential steps. What these abbreviated inferences may best be, will depend upon the special nature of the subject under discussion. But in my algebras and graphs, far from anything of that sort being attempted, the whole effort has been to dissect the operations of inference into as many distinct steps as possible.[49]

Interestingly, in this quote Peirce seems aware of the tension which exists between building a system with which to perform useful work, and building a system with which to explore the nature of logical inference. Peirce places his system of existential graphs squarely in the latter category, and this again very likely contributed to the cool reception his system has received.

Yet, as we have suggested, there were also deeper, more foundational reasons why Peirce's graphs did not have an impact on the practice of logic of the time. These have to do with the grounds one might have for adopting existential graphs, given all of the disadvantages we have just outlined. In order to understand them, we will need to briefly describe the intellectual context in which Peirce conceived his graphs, and the function which he intended them to fill within his overall philosophy. Unfortunately, the full setting of Peirce's logical work is extremely complicated, and we cannot do it justice here. However, our explanation will be sufficient to answer why, given all of Peirce's disclaimers for his system of existential graphs, he spent the last ten years of his life working on it, and considered it his greatest logical achievement. After all, Peirce is quite clear that his reasons for his work on existential graphs are not related to their aesthetic qualities or ease of use:

> [A]lthough there is a certain fascination about these graphs, and the way they work is pretty enough, yet the system is not intended for a plaything, as logical algebra has sometimes been made, but has a very serious purpose...[50]

---

[49] Peirce 1931, 4.424. In his emphasis in providing a system which dissects inference into "as many distinct steps as possible," Peirce is echoing a theme sounded by Frege in Frege 1879.

[50] Peirce 1931, 4.424.

Let us investigate the "serious purpose" which Peirce had in mind for existential graphs.

The importance of existential graphs in Peirce's philosophy arises from their central function at the intersection of his general theories of logic, semiotics, and philosophy of science. As a result of his underlying theory of pragmatism, Peirce believed that the process of abstract reasoning, of working with signs, was at root the same as the process of scientific experimentation. In particular, he held that the sorts of objects upon which reasoning takes place should be constructed in such a way as to make the reasoning process as similar as possible to a more traditional observational science. Existential graphs were designed to fill that role: they would make both the logical structure and the entailments of propositions directly observable, in the same way that (as we have earlier observed) he believed that molecular diagrams make both the atomic structure and possible combinations of organic compounds observable. This theme of Peirce's logical work – that reasoning is a type of semiotic; and hence involves a process of assigning meaning to physical signs; and hence should include an experimental methodology of manipulating its signs and directly observing the consequences of this manipulation – is at the heart of Peirce's argument for the importance of existential graphs.[51]

As a result of this view, Peirce's definition of the types of signs which can constitute an existential graph involves an important restriction. This restriction, which is based on the overall classification of signs in Peirce's semiotic theory, limits the types of signs present in existential graphs to icons and indices, and in particular places very heavy restrictions on signs of the sort Peirce calls "symbols."

> A *diagram* is a representamen which is predominately an icon of relations and is aided to be so by conventions. Indicies are also more or less used.[52]

This restriction secures for existential graphs at least three properties which are important for Peirce's desired experimental character of reasoning.

---

[51] See, *e.g.*, quotes like "Logic, in its general sense, is, as I believe I have shown, only another name for semiotic" (Peirce 1931, 2.227).

[52] Peirce 1931, 4.418. In existential graphs, symbols are used only to name the base objects and classes, and never to designate logical relations. Peirce's division of signs into icon, index, and symbol (and their respective subtypes) is extremely complex, and space limits its explication here. See, *e.g.*, Peirce 1940.

First, the restriction to icons and indices guarantees the logical possibility of the referents of the diagram's signs:

> But there is one assurance that the Icon does afford in the highest degree, namely, that which is displayed before the mind's gaze – the Form of the Icon, which is also its object – must be logically possible.[53]

Second, Peirce perceived clearly that reasoning with icons and indices allows us to offload much of the symbolic reasoning task to the physical structure of the icons. Because icons and indices are defined relative to the structural properties which they share with their referents, reasoning with these sorts of signs can take advantage of these properties. Peirce thought that reasoning with his graphs would have the character of repeated cycles of manipulation and observation over the graphs, where the properties of the icons relative to the manipulations guarantee that the reasoning is appropriate for the objects represented by the icons. According to Peirce, the advantage of such an approach to reasoning may not be apparent with very simple logical systems, but will become obvious when the reasoning task is expanded to include the logic of relations:

> Deductive logic can really not be understood without the study of the logic of relatives, which corrects innumerable serious errors into which not merely logicians, but people who never opened a logic book, fall from confining their attention to non-relative logic. One such error is that demonstrative reasoning is something altogether unlike observation. But the intricate forms of inference of relative logic call for such studied scrutiny of the representations of the facts, which representations are of an iconic kind, in that they *represent relations in the fact by analogous relations in the representation*, that we cannot fail to remark that it is *by observation of diagrams that the reasoning proceeds* in such cases. We successively simplify them and are always able to remark that such observation is required, and that it is even thus, and not otherwise, that the conclusion of a simple syllogism is seen to follow from its premises.[54]

Finally, Peirce also believed that the use of a graphical system built of icons and indices would bring out the exploratory character of experimental reasoning. Unlike reasoning with sentential systems, whose use suggests that only a single conclusion can be drawn from a set of premises, Peirce de-

[53] Peirce 1931, 4.533.

[54] Peirce 1931, 3.641, emphasis added.

signed his existential graphs to simultaneously support multiple possibilities and conclusions, each one latent in the structure of the iconic representation until it is brought forth by the observations of the reasoner:

> [N]on-relative logic has given logicians the idea that deductive inference was a following out of a rigid rule, so that machines have been constructed to draw conclusions. But this conception is not borne out by relative logic. People commonly talk of the conclusion from a pair of premises, as if there were but one inference to be drawn. But relative logic shows that from any proposition whatever, without a second, an endless series of necessary consequences can be deduced; and it very frequently happens that a number of distinct lines of inference may be taken, none leading into another. That this must be the case is indeed evident without going into the logic of relatives, from the vast multitude of theorems deducible from the few incomplex premises of the theory of numbers. But ordinary logic has nothing but barren sorites to explain how this can be.[55]

For Peirce, these three features of iconic reasoning form a decisive argument for the importance and value of such representational systems. And, from our point of view, we can see that Peirce's case for the importance of existential graphs rests on an entirely different sort of argument for the desirability of diagrammatic representations in logic than we have previously seen, and indeed, an argument which is in many ways strikingly similar to claims which motivate the Hyperproof project today.[56]

Unfortunately, it seems that the radical nature of Peirce's critique of traditional systems of logic caused his system of existential graphs to be marginalized within the community of practicing logicians, especially those in Europe. During the decades at the close of the nineteenth and beginning of the twentieth century, virtually all of the influential work in logic was done by European-based logicians working in an extremely mathematical framework, such as Frege, Russell, Peano, Hilbert, and Skolem. Peirce's ideas about the experimental nature of actual reasoning never attracted a following among these logicians, and seem to have had no effect on the European mainstream of logic. For example, Peirce is cited less than five times, and then only for his work in the logic of relations, in the forty-six seminal papers in logic collected in van Heijenoort's famous *Sourcebook*.[57] Also, it must be noted that Peirce's work habits did not help his popularity. Although his output was prodigious (10,000 pages published during his life-

[55] Peirce 1931, 3.641.

[56] See Barwise and Etchemendy 1994.

[57] van Heijenoort 1967.

time, plus a total of roughly 100,000 pages of unpublished manuscript), it was disorganized and often only partially complete. Kneale and Kneale remark that:

> Unfortunately Peirce was like Leibniz, not only in his originality as a logician, but also in his constitutional inability to finish the many projects he conceived.[58]

Without an understanding and acceptance of Peirce's philosophical position, which (even had they known about it) more mainstream logicians like Frege and Hilbert would have been unlikely to give, Peirce's existential graphs offered no compelling advantage over the traditional systems of notation, and indeed were more cumbersome in several ways. For this fundamental reason, I believe, Peirce's existential graphs never became part of the dominant flow of European logical research at the turn of the century, and remain now only a footnote in the history of logic.

10.5 Logic at the End of the Nineteenth Century

The final section of our investigation of diagrams in symbolic logic will be concerned with some observations about the developments in logic which occurred in Europe from Frege's 1879 publication of his *Begriffsschrift* to the work of Hilbert in the early 1900s. Here, at last, is the point at which the two major histories which have concerned us here begin to come together. Recall that Frege and Hilbert played a significant role in our account of diagrammatic reasoning in late nineteenth century geometry with which we concluded Part I. Frege's logical work was done in the context of a philosophical position which was an interesting transitional point between traditional universalistic views of logic and the emerging more limited conception. Certainly, Frege accepted a version of the established Aristotelian/Kantian doctrine of *logica magna* – the principle that the discipline of logic was properly thought of as a single universal theory encompassing a set of perfectly general laws which govern entailment relationships of all kinds. His work was explicitly concerned with deriving arithmetic from a preexisting set of universal logical principles which could be known *a priori*, and in this way securing a necessary basis for the truth of statements of mathematics. However, Frege's intense focus on arithmetic as an applica-

---

[58] Kneale and Kneale 1962, pg. 427.

tion domain, to the exclusion of other, more traditional logical concerns (such as the nature of modality), also made his work a natural forerunner to Hilbert's attitude of *logica utens* – the idea that logic was not a monolithic collection of the eternal laws of thought, but rather a set of distinct theories connected more through a similarity of approach than by any specific laws or principles.[59] The flexibility and relative freedom from large-scale metaphysical commitments which *logica utens* promised was certainly part of the attraction of the later philosophies of logic which embraced it, such as Hilbert's logical formalism. Interestingly, as the growing acceptance of *logica utens* as a paradigm for research made the notion of a specialized logic become more and more legitimate, it seems that a natural consequence would be that diagrammatic systems of logic would finally emerge and be research topics in their own right. The reasons why such systems never appeared are closely related to the reasons we described earlier for the final decline of diagrammatic methods in geometry, and will be the topic of this final section.

We can suggest two main reasons for the lack of development of diagrammatic systems in mainstream logic between 1880 and 1900, one related to the technical developments in logic which happened over this period, and one related to changes in its philosophical background. We shall examine the technical developments first. Spurred by the attention paid to logic by several of the leading mathematicians of the time, symbolic logic underwent a very rapid evolution over those twenty years. Our previous discussion of geometry has already touched on the fundamental changes that swept through mathematics during the middle part of the nineteenth century. These changes included an increasing concern for the abstract structure of mathematical entities, a wholesale reevaluation of the foundations of algebra and analysis, and a strong program of rigor in mathematical proof. Along with Peirce, Venn, and De Morgan, Frege was one of the early logician-mathematicians who used Boole's ideas as a starting point from which to apply these new systematic concerns to the logical underpinnings of mathematics. Frege's research program was to use the framework of symbolic

[59] This "similarity of approach," which Hilbert used to conceptually unify the different deductive systems which function in different areas of mathematics, had several characteristics. First, although Hilbert apparently did not accept different interpretations for the logical constants (such as "¬" and "="), he required complete freedom in interpreting the nonlogical terms and axioms. He also allowed different logics to have different languages, and therefore to be formally incompatible because they might place different structural requirements on the same nonlogical term. However, Hilbert explicitly rejected logical intuitionism, and held that the possible rules of inference in proofs could vary only inasmuch as they would remain checkable using only the cognitive facilities available under the *finite Einstellung*, which we will describe below.

logic to show that arithmetical reasoning and concepts could be reduced to logical reasoning and concepts, and further that the logical constructs which were necessary to this reductionist program were constructible from those which were *a priori* available to us as a consequence of the very possibility of rational thought itself. Frege immediately saw that pure Boolean logic was unsatisfactory for this task, and he radically modified Boole's original framework in several ways. One of the most important was the introduction of the function/argument formulation of predicate logic. Boole, as we have observed, replaced the classical subject/predicate distinction in logic with a system based on relations between class terms, and had even occasionally used functions in place of these terms, but it was only with Frege's work that the formalism of functions and arguments achieved the central place it holds today. And, through the investigation of the suitability of this new function-based logic for mathematical statements, Frege was led to his second major innovation: that of quantification theory and the notion of the bound variable.

Frege proposed all of these changes in his 1879 *Begriffsschrift*, a book whose very title signals the reader that his purpose is to present a new system of expression for logic.[60] It is worth remembering that Frege probably felt that, at the time of the *Begriffsschrift*, the magnitude of the changes to logic that he was proposing justified the introduction of a new kind of notation which would better communicate his ideas. And, although Venn claimed that Frege's unique notation was developed "in apparent ignorance that anything better of the kind had ever been attempted before," it is not clear that any other system of notation existing in the 1870s would have been able to easily satisfy all of Frege's criteria.[61] Besides being able to accurately express his new notions of function, quantifier, and variable scope, Frege also required his notation, or "concept-script," to support his goals for a proof theory as well:

> Its first purpose, therefore, is to provide us with the most reliable test of the validity of a chain of inferences and to point out every presupposition that tries to sneak in unnoticed, so that its origin can be investigated.[62]

We should note that, in his call for deductive clarity, Frege's objective for his notation is quite similar to Peirce's, although Frege certainly did not

[60] Recall that the full title of Frege 1879 is, "*Begriffsschrift*, A Formula Language, Modeled Upon That of Arithmetic, For Pure Thought."

[61] Venn 1884, pg. 493.

[62] Frege 1879, pg. 6.

subscribe to Peirce's further goal that a notation also be designed to turn the process of proof into an observational science.

In addition to the expressive and proof-theoretic requirements that his domain of mathematical statements placed on his notation, Frege also had larger ambitions for it. Because of the way that the basic concepts of function, quantifier, and argument are integrated into the structure of his notation, Frege hoped that the use of his new "ideography" would be a valuable tool in solving more general problems in philosophy:

> If it is one of the tasks of philosophy to break the domination of the word over the human spirit by laying bare the misconceptions that through the use of language often almost unavoidably arise concerning the relations between concepts and by freeing thought from that with which only the means of expression of ordinary language, constituted as they are, saddle it, then my ideography, further developed for these purposes, can become a useful tool for the philosopher. .... The mere invention of this ideography has, it seems to me, advanced logic.[63]

Interestingly, although Frege's holistic view for his logical system was similar to that of Boole, the specific goals which Frege had for his concept-script are distinct and in a sense opposite from Boole's. While Boolean notation and rules were structured to emphasize the similarities between logic and algebra, Frege's foundational project for arithmetic required that he keep his logical notation formally distinct from the mathematical symbols of his intended subject matter, or of any other possible subject matter. Frege acknowledged as much in an 1882 letter to Schröder:

> My intention was not to represent an abstract logic in formulas, but to express a content through written signs in a more precise and clear way than it is possible to do through words. In fact, what I wanted to create was not a mere *calculus ratiocinator* but a *lingua characterica* in Leibniz's sense.[64]

Frege was fully aware of the implications of his reference to Leibniz's *lingua characterica*. He conceived of a logical system which was completely independent of subject matter, including only those laws of combination which would be valid on arbitrary concepts. These laws, the "laws of pure

---

[63] Frege 1879, pg. 7.

[64] See van Heijenoort 1967, pg. 2.

thought," were to be the base from which he hoped to derive the basic principles of arithmetic. In this way, he aspired to show that the edifice of mathematics was a consequence of these laws, and therefore that the theorems of mathematics were necessary consequences of the structure of rational thought itself. For this project to be successful, however, the operations of the underlying logic had to be completely universal and able to be applied to any concept which could be the object of rational thought. Therefore, the notation which Frege devised for this logic, the concept-script, had to be equally general.

The system of expression which Frege set out in the *Begriffsschrift* seemed to Venn to have a character somewhere between Boolean notation and Venn diagrams. Indeed, Venn wrote that it "deserves to be called diagrammatic almost as much as symbolic."[65] Unlike the diagrams of Euler or Venn, however, Frege's notation does not employ bounded areas as representatives for classes; rather, it uses arcs and lines of a particular sort to indicate the propositional and quantificational relations which exist between sententially specified variables and subformulas. For example, Frege would express the Aristotelian categorical sentence, "Some P are M," in the following way:[66]

Figure 17: Frege's Notation for "Some P are M."

Complex conditionals can be written as single judgments encompassing both the antecedent(s) and the consequent, in the following example of one of Frege's "judgments of pure thought," or axioms of logic:

65 Venn 1884, pg. 493.

66 I will use bold letters (*e.g.*, "**a**") in place of Frege's German script.

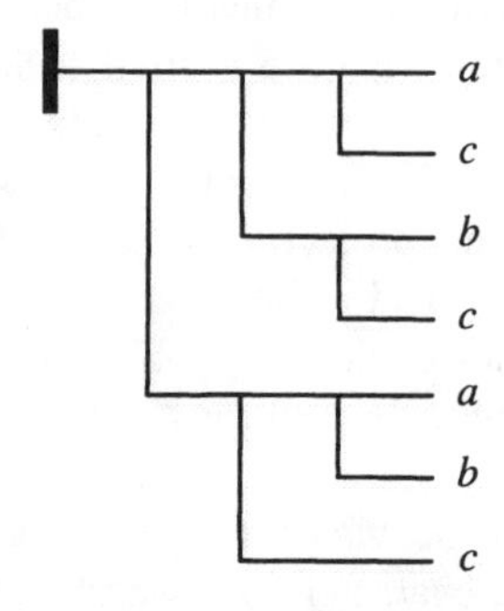

Figure 18: A Complex Conditional in Frege's Notation

This expression represents the propositional judgment that: "if *a* follows from *b* and *c*, and *b* follows from *c* alone, then *a* must also follow from *c* alone." We can see that each of the propositional subcomponents has a place in Frege's representation of it. Frege's notation also has a number of features which allow it to represent specialized expressions in mathematics, as are exhibited in the following more complex example:

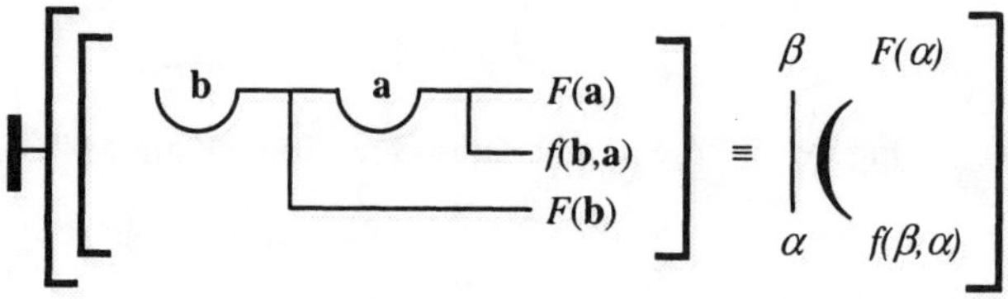

Figure 19: A Complicated Frege Diagram

This expression can be interpreted as: "if from the proposition that **b** has property *F* it can be inferred generally, whatever **b** may be, that every result of an application of the procedure *f* to **b** has property *F*, then I say: property *F* is hereditary in the *f*-sequence."[67]

Frege gave the system of the *Begriffsschrift* only a single rule of inference, based on the principle of *modus ponens*, or detachment:

Figure 20: *Modus Ponens* in Frege's Notation

With the appropriate substitutions, Frege asserts that this rule will be sufficient to allow his system to encompass all of the traditional Aristotelian syllogistic inference forms. He suggests using the following strategy:

> Following Aristotle, we can enumerate quite a few modes of inference in logic; I employ only this one, at least in all cases in which a new judgment is derived from more than a single one. For the truth contained in some other kind of inference can be stated in one judgment, of the form: if M holds and N holds, then L holds also, or, in signs,

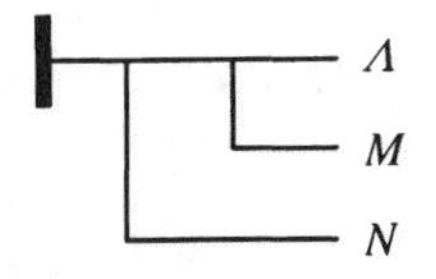

> From this judgment, together with:

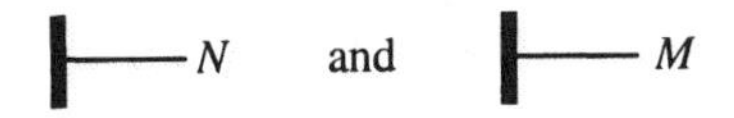

> there follows, as above:

[67] Frege 1879, pg. 57.

> In this way an inference in accordance with any mode of inference can be reduced to our case. Since it is therefore possible to manage with a single mode of inference, it is a commandment of perspicuity to do so.[68]

However, *contra* this implication that it will be more perspicuous to use only this rule of inference, Frege also explicitly acknowledges that he does not intend for his restriction to *modus ponens* to carry any psychological weight, and claims that he structured his system in this way merely because it was expedient to do so.

In spite of the advantages of clarity and precision which Frege claimed for it, the community of logicians never adopted the concept-script of the *Begriffsschrift*. One obvious reason for this arises from the significant practical drawbacks involved in actually using Frege's notation. Even employing the various abbreviations proposed by Frege in the *Begriffsschrift*, the authoring of individual propositions and proofs using the concept-script requires large amounts of space on a page and (because of the two-dimensional layout and number of alphabets which Frege employs) results in complications when publishing the results. In response to these issues, Frege rather tersely pointed out that the "comfort of the typesetter is certainly not the *summum bonum*," but this tradeoff was apparently not one which his fellow logicians were willing to make.[69] Venn, for example, criticizes the "inordinate amount of space" demanded by the concept-script, complaining that, "nearly half a page is sometimes expended on an implication which, with any reasonable notation, could be compressed into a single line."[70] Kneale and Kneale refer to the "forbidding appearance" of Frege's symbolism and point out that even Russell claimed that the *Begriffsschrift* notation made it impossible to understand or even grasp the importance of Frege's work on a first reading.[71] By 1894, Peano had introduced a notation for symbolic logic which retained the syntactic precision and expressiveness of Frege's concept-script, but which was substantially more compact and easier to print. Peano's notation was subsequently adopted by Hilbert, and was refined by Russell and Whitehead for use in *Principia Mathematica*. The typographical advantages of Peano's notation, and its accep-

---

[68] Frege 1879, pg. 17.
[69] See van Heijenoort 1967, pg 2.
[70] Venn 1884, pg. 494.
[71] Kneale and Kneale 1962, pg. 511.

tance by the most prominent logicians of the time, certainly helped to insure that Frege's concept-script did not survive.

This brief history of Frege's concept-script illustrates one of the reasons we suggested above why other diagrammatic or quasi-diagrammatic systems for logic were not developed during this otherwise extremely fertile period in logic. With the exception of Peirce's existential graphs, whose construction was motivated by a unique set of philosophical concerns, no other diagrammatic system for logic emerged during this time which was capable of handling the increased expressive and proof-theoretic requirements of the new quantificational logics. This means that, in spite of the obvious disadvantages of Peano-style notation (*e.g.*, its strictly linear organization, which makes quantifier scoping and subformula grouping difficult to see), any competing diagrammatic notation which was conceived by logicians during this time must have been believed to involve unacceptable sacrifices of space or convenience. Put another way, no general-purpose diagrammatic system was developed during this period (or, in fact, has been since) whose technical or cognitive advantages over the standard notation were sufficiently compelling to spur its widespread adoption, or to even make a significant impact on logical practice. Recent investigations of the use of diagrams in formal systems have concentrated on explaining this fact by appealing to the inherent expressive limitations of "homomorphic" representations, such as their apparent inability to naturally represent general negation or disjunction.[72] These investigations essentially confirm that the increase in the sophistication of symbolic logic's representational demands, caused by the work of Frege and others, simply overwhelmed the ability of simple diagrammatic systems to function as convenient and cognitively effective representations for general logic. Basically, because of the additional technical requirements of quantificational logic, simple diagrammatic systems were no longer representationally adequate, and diagrammatic systems that were powerful and complex enough to meet these requirements held no compelling advantages over the existing Peano-style notations.

Besides this technical reason, however, there was also a deeper and more philosophical reason for the general lack of interest in any sort of logical system which exploited geometric relationships beyond those used in the accepted notations for mathematical formulae. This reason was rooted in the intense distrust of intuition which permeated mathematics and was embraced by many of the leading mathematicians of the time. In chapter 5, we described this attitude in the context of nineteenth-century geometry, and discussed how it influenced the geometrical work of Pasch and Hilbert.

---

[72] See, *e.g*,, Barwise and Etchemendy 1991.

Frege, too, held a variety of this view, as a result of his commitment to the purity and reliability of the laws of thought which lay at the heart of his logicism. His philosophical project for mathematics required that these laws be completely analytic, and not be linked to our experiences of space or time or any other presumably synthetic concept. Hence, because geometry is the study with space and extension, Frege believed that its theorems would be necessarily dependent on this type of intuition:

> There is ... a noteworthy difference between geometry and arithmetic in the way in which their fundamental principles are grounded. The elements of all geometrical constructions are intuitions, and geometry refers to intuitions as the source of its axioms. Since the object of arithmetic does not have an intuitive character, its fundamental propositions cannot stem from intuition either.[73]

This conviction about the intrinsically intuitive nature of geometric objects did not prevent Frege from speculating that once a set of (intuitively-justified) axioms was supplied, an adequately rigorous proof system could be given for geometry. In fact, he suggested that such a geometric proof system could be built on his existing *Begriffsschrift* notation. Nevertheless, geometry itself could never be *a priori* in the way that arithmetic can, because of the *a posteriori* character of its base concepts.

> It seems to me to be easier still to extend the domain of this formula language to include geometry. We would only have to add a few signs for the intuitive relations that occur there. In this way we would obtain a kind of *analysis situs*.[74]

The mechanisms of expression and proof in the *Begriffsschrift*, Frege thought, were constructed to be entirely free of the contaminating influence of intuition; it was because of this that they could constitute a reliable formalism for geometric reasoning. Frege believed that the formal reliability of his system would be guaranteed by the care he had taken in defining its syntax and inference rules so that they would be immediately reflective of his abstract and universal laws of thought. From a modern standpoint, it is not clear that Frege can coherently maintain this position, in view of the fact that the actual interpretation of his concept-script depends on relationships

---

[73] Frege 1984.

[74] Frege 1879, pg. 7.

which are expressed via the geometric relations of lines, arcs, and subformulas. However, it is certain that Frege would have viewed the diagrammatic systems of Venn and Peirce as unacceptably dependent on irreducibly geometric notions, such as interior and incidence, whose roots he felt must lie in intuition.

The antipsychologism which Frege brought to logic was strengthened and reinforced in the work of Hilbert. In addition to his immense prestige as a geometer, Hilbert was also recognized as one of the foremost logicians of his time. Hilbert's commitment to axiomatization, rigorous deduction, and independence from subject matter in the foundations of geometry was mirrored in his approach to logic. In fact, Hilbert's concern with logic and the foundations of mathematics stemmed directly from his experiences in crafting axiomatic systems for pure geometry. Hilbert came to believe that the various different disciplines of mathematics all shared a common deductive nature, and were best understood as following out the consequences of different axiom sets which shared a collective core. He stressed this unified character in a famous essay written soon after the publication of the *Foundations of Geometry*:

> [I]n my view, mathematics is an indivisible whole, an organism whose ability to live is governed by the connection between its parts. No matter how various is the stuff of mathematical knowledge in its details, we recognize very clearly the sameness of the logical tools, the close relations between idea formation across the whole of mathematics, and the innumerable analogies between its distinct domains.[75]

Hilbert's reference here to the "sameness of logical tools" is an early indication of his conviction that logic and the axiomatic method played a central role in mathematics. His work on clarifying the status of mathematical theories and the role of these logical tools within them gave rise to a philosophy of logic which is substantially different from the Fregean account, and whose influence is easily discernable today.

It will be helpful to briefly examine the major differences in philosophy of logic between Hilbert and Frege. We described earlier how Frege held that the axioms from which mathematics was to be ultimately derived, the fundamental laws of pure rational thought, had to be universal, analytic, and entirely free of subject matter. In contrast, Hilbert flatly rejected the notion that the logic operative in mathematics had to be taken directly from this hypothesized universal framework of rational thought, whose existence and

[75] Hilbert 1900, pg. 329.

properties would be necessarily prior to the concerns of mathematics. As we shall see, Hilbert's project will require the ability to use logical and mathematical techniques to investigate the properties of mathematical systems, and because logic was a critical part of these systems, the structure of logic itself could not be thought of as prior to study without inviting circularity. Further, Hilbert also disagreed with Frege's account of the reference of terms used to denote arithmetic objects. Frege held that these terms (the numbers) simply referred to specially-constructed extensions of a certain sort. Hilbert objected to this, claiming that the interpretation of the terms in any of the axioms of mathematics could not be limited in advance to a particular type of referent or subject matter. As with the axioms of geometry, Hilbert held that their basic referring expressions should have *no* privileged domain of interpretation, but, rather, should be explicitly neutral between any of the possible domains that possess the correct structural properties. However, for the case of the basic axioms of arithmetic, Hilbert pointed out that when the referents are interpreted as being about counting relationships between extralogical signs of a certain primitive type (marks of the form |, ||, |||, etc.), then the statements of mathematics can be assigned enough content so that we can intuitively judge them to be true or false without the requirement of formal mathematical proof. Even in this case, though, there would be no necessary connection between the terms of the mathematical axioms and the objects in this domain of marks, because there is no requirement that the axioms be interpreted in this way.

These attitudes about the structure of mathematical axioms and the role of logic in mathematics led Hilbert to the belief that the foundations of mathematics could be secured if the following three conditions were met: that all the axioms of mathematics were constructed so as to be free of all subject matter, and therefore reference-neutral; that the axiom sets for each subfield of mathematics were themselves consistent, complete, and independent of each other; and that the proofs which proceed from these axioms can be rigorously shown to be trustworthy. Hilbert's approach to mathematics can be understood as his attempt to build a foundational structure which conforms to each of these conditions. For our purposes, the first and third of these also have significant implications for the possibility of diagrammatic systems in logic, and so with this in mind we will address these in turn.

First, as we have observed, Hilbert required that the axiom systems used in mathematics be completely neutral and reference-free. We have already described in chapter 5 Hilbert's grounds for requiring that geometric axiomatic systems have this property, and for our present discussion it will suffice to say that Hilbert adopted his reasoning from the geometry domain into his new concern with the more general logical foundations of mathematics. Recall that in chapter 5 we also showed that this position entailed that Hilbert could not accept any sort of diagrammatically-based methods in geometry. These sorts of arguments still apply in the larger context of the general foundations of mathematics and logic. Diagrammatic logical systems like Peirce's or Venn's would have been immediately problematic for Hilbert, because the very statement of the axioms using these sorts of systems must be done graphically, and therefore must entail a particular, probably classically Euclidean, interpretation of the meaning of the critical geometric properties, such as "inside" and "connected." This sort of interpretation is no different from the interpretations required by diagrammatic methods in geometry, which we have already shown that Hilbert must reject because of its conflict with his reference-free view.[76] Beyond this issue, however, lies a deeper conflict between Hilbert's formalist philosophy and the use of diagrammatic systems in mathematical reasoning. Since Hilbert insisted that the axioms and statements of mathematics must be neutral with regard to any particular type of referents, then this very absence of a privileged domain of interpretation entailed that the symbol strings of the axioms *themselves* become what is meant by a mathematical theory. For example, in this view the role of Hilbert's axioms for Euclidean geometry would not be to express truths about any logically prior Euclidean domain; rather, they *themselves* are what is meant by Euclidean geometry. So, if the axioms for a given part of mathematics are essentially graphical, as would happen in a diagrammatic system of logic such as Peirce's, then by the arguments of chapter 5 their use can be no more free of unintended interpretations or unarticulated assumptions than the original Euclidean diagrams for geometry. On these grounds, then, Hilbert would have been forced to reject a diagrammatically-based logical system like Peirce's or Venn's.

The second way that Hilbert's philosophy of mathematics impacts the possibility of diagrammatic logical systems is through the demands of his metamathematical program for proofs. We have touched on this argument in chapter 5 as well. In order to achieve an adequate standard of rigor for proofs in his system, Hilbert required that the proof procedures used in mathematics and logic conform to certain standards of finiteness and exactness. In his early work on the axiomatic systems used in geometry, he sim-

[76] Of course, Hilbert's work in the *Foundations of Geometry* (Hilbert 1899) provides a way to describe these sorts of geometric properties in a formally acceptable way, but appeal to this sort of description as the basis for a diagrammatic deductive system would simply amount to a translation of that system into a logically prior sentential one which uses the geometric vocabulary of the *Foundations*.

ply posited that these requirements came from the "general philosophical requirements of reason:"

> ...above all, I stress that a solution [to a mathematical problem] is to be achieved by giving the answer through a finite number of inferences, and indeed based on a finite number of assumptions, assumptions which are involved in the very posing of problems, and which can be precisely formulated. This demand .... corresponds to a general philosophical requirement of reason. Indeed, only through its satisfaction are the conceptual content and fruitfulness of the problem fully clear.[77]

Hilbert's later work in logic, however, significantly expanded and clarified this basic view. By 1905, Hilbert had specified that the sorts of discrete signs which could appear in mathematical expressions must be drawn from a finite alphabet, and that the rules of composition for these signs are required to be patterned after the rules of elementary arithmetic:

> [T]he role of the language in the expression of the logical connections between thoughts corresponds to the sign language in calculation. In following a logical passage of thought with the help of this logical language, we carry out simultaneously a calculation, in which manifold logically elementary processes are put together according to practiced rules. It is even self-evident that, when we exclude the accidental features in the derivation of words, then a form of mathematical sign language arises.[78]

Hilbert's repeated references to carrying out a "calculation" signal that he intends that the process of mathematical proof over these sorts of expressions will be akin to the combinatorial, mechanical processes of arithmetic. And, while Hilbert is never precise about the exact parameters of this calculating processes, he is clear that he intends that the carrying out of this calculation (and thus, the application of the rules of proof in mathematics) should require only the ability to determine whether or not a particular symbol string is syntactically identical to another. Hilbert understood that this recognition ability is basic to our ability to use symbolic systems, and in an unpublished lecture further suggested that it is a prerequisite to coherent thought:

---

[77] Hilbert 1900, pg. 293.

[78] Translated in Hallett 1994, pg. 180. See also our analysis of this attitude in our discussion of Hilbert's philosophy of geometry in Part I.

> ... an Axiom of Thought, or, as one might say, an Axiom of the Existence of an Intelligence, which can be formulated approximately as follows: I have the capability to think things and to denote them through simple signs (*a*, *b*; ..., *X*, *Y*, ...; ...) in such a fully characteristic way that I can always unequivocally recognize them again. My thinking operates with these things in this designation in a certain way according to determinate laws, and I am capable of learning these laws through self-observation, and of describing them completely.[79]

By the early 1920's, however, Hilbert realized that limiting the process of proof to only those operations which could be described by a calculating process resulted in proofs which themselves could be the objects of a special kind of mathematical investigation. Further, he also saw that this requirement entailed that proofs were of finite length, and hence that the mathematical investigation of the metamathematical properties of the axiom system could be carried out using only finitary means, even though the actual proofs under study might be concerned with transfinite inferences. And, because of his formalist view that (as we observed above) different mathematical subfields are nothing more than the collection of symbol strings and proof procedures which are associated with them, this kind of metamathematical investigation would not merely be an investigation of a particular symbol system, but rather it would be an investigation of the metamathematical properties of the chosen mathematical subfield itself.

Therefore, in order for Hilbert to accept diagrammatic proofs as a legitimate method of logical inference in mathematics, he would have needed to accept that they were able to be the sorts of objects which could be analyzed by finitary methods alone. This is because the restriction to finitary techniques in metamathematics which we have just described was critical to Hilbert's subsequent logical project. In order for Hilbert to avoid a regress in his account of the role of logic, he proposed that the ability to declare that a particular proof contained no mistakes in the application of its rules, and was therefore valid, would rest on our ability to simply and immediately intuit that all the necessary logical rules were correctly applied. The very weak form of intuition to which Hilbert appeals here is simply the above-mentioned ability to calculate with signs, or perform elementary arithmetic. In this way, he hoped to ground his philosophy of mathematics in a kind of logical intuition which would be rudimentary enough that it would be safe

[79] Translated in Hallett 1994, pg. 179.

from the sorts of problems in naive logical intuition which Russell discovered in Frege's account. As Hilbert says concerning this ability:

> ...in elementary domains of arithmetic, in particular that of elementary number theory, there is that complete certainty in our considerations. Here we get by without axioms, and the inferences have the character of the *tangibly certain.*[80]

Our inherent ability to recognize truths using this "finite point of view," or *finite Einstellung*, is the mechanism which provides the ultimate guarantee in Hilbert's philosophy that the proof systems which are used in mathematics are consistent, and therefore that the mathematical systems themselves are acceptable. Hence, the legitimacy of diagrammatic proof systems in Hilbert's philosophy will be dependent on whether such systems give rise to proofs which are amenable to analysis and intuitive validity judgments under the *finite Einstellung*.

We can judge this by looking at an example of the sort of domain in which Hilbert believes we can immediately intuit the truth of assertions. For the facts of elementary arithmetic, Hilbert suggests a translation into a sign-counting domain.[81] For example, let us assign the arithmetic sign 1 to the syntactic object "|", 2 to "||", and so on; and let "+" be the (intuitive) concatenation operation and "=" be the quantitative identity operation. Then, the claim that "||" concatenated with "|||" is identical in number of strokes to "|||||" would an interpretation of "2+3=5" in a domain in which Hilbert holds that we can immediately apprehend the truth of claims. Interestingly, when Hilbert's collaborator Paul Bernays attempted to explain this point about the ability of simple signs like these to support immediate truth judgments, he suggested a distinctly Peircean explanation. He held that reasoning with these signs essentially depends on their ability to physically capture the relevant features of their referents:

> The philosopher is inclined to speak of this representation [between sign and number] as a relation of meaning. However, one should note that, in contrast to the usual relation between word and meaning, there is [in this example] the essential difference that *the object doing the representing contains the essential properties of the object to be represented.* Thus, the relations which are to be investigated between the objects represented *are*

[80] Translated in Hallett 1994, pg. 184, emphasis mine.
[81] See Hilbert 1904.

> *to be found in* the objects doing the representing, and thus can be established through consideration of these.[82]

Certain kinds of diagrams, of course, can also make use of this sort of physical encoding of properties, as Peirce demonstrated. Nevertheless, given Hilbert's view on the unreliability of intuition in geometric proof, it is relatively certain that Hilbert would have thought that the sorts of spatial intuition which would allow us to interpret sophisticated logic diagrams would not be sufficiently trustworthy to admit into the stock of techniques of metamathematics. Because of this, we can conclude that inference systems which rely on geometric relations for their interpretation, such as the diagrammatic systems of Peirce or Venn, would not have been accepted by Hilbert as a legitimate deductive technique in logic. And, although Hilbert may not have been entirely justified in this position, especially given the recent work in diagrammatic systems described in Allwein and Barwise 1986, I believe that considerations of this sort can adequately explain the lack of development of diagrammatic systems in logic in the twentieth century.

[82] Translated in Hallett 1994, pg. 185, emphasis mine.

11

Summary

Let us review what we have accomplished in Part II. We began by wondering why the formal analysis of correctness in reasoning has, almost uniformly throughout the history of logic, reduced to the analysis of sequences of sentences in some language. Our primary strategy in addressing this question was similar to the strategy we employed in our investigation of diagrammatic methods in geometry: to look for answers in the intellectual history of the discipline, concentrating on the period from Aristotle's initial work to the turn-of-the-century accomplishments of Hilbert. By tracing the evolution of the typical sorts of problems which logic has addressed, and by examining the techniques which were developed to solve them, we have been able to show how the linguistic techniques which were originally created for reasoning in a syllogistic framework have shaped the standards for more modern systems of logical inference. More interestingly, though, we have been able to identify *reasons* for the lack of development of diagrammatic methods in logic. These reasons are rooted in the influence of certain beliefs about the background metaphysical context and appropriate subject matter which were present at the very beginnings of logic, and which have persisted and affected the structure of succeeding logical theories. The main accomplishment of this part of the book has been to describe the effects of these ideas on the subsequent development of logic, and in this way argue that the lack of diagrammatic methods in logic today is not due to lack of creativity, but rather is a consequence of a set of specific intellectual aims which guided the evolution of logic.

As has been made clear in this part of the book, a single fundamental principle has been at the center of the way that logicians from Aristotle to Frege have structured their accounts – namely, that the scope of a legitimate logical theory should be as broad and general as possible. Let us call this the *principle of maximal scope*. By this, it is not intended that logic should capture *all* types of reasoning; for example, the function of classical logical theory in reasoning about modal claims, or probabilistic claims, or claims involving complex quantitative relations, has been difficult and controversial for most of logic's history, and providing a formal account of such reasoning has usually been accepted as secondary to providing an account of the entailment relations between simple categorical statements. Rather, the principle of maximal scope entails that logic should not be artificially limited in its domain of applicability, and thus it should attempt to model whatever is common about reasoning broadly conceived, however small that common fraction may be. Logicians working in this *logica magna* tradition have always perceived logical theories which are intrinsically restricted to analyzing only reasoning in biology or taxonomy or some other individual discipline as possessing a severe shortcoming, regardless of how effective the theory may be at modeling valid reasoning in the intended domain. Following the principle of maximal scope, the subject matter of the most influential logical theories has typically been defined so that these theories can claim to capture that portion of valid reasoning which is common across domains as extensive as all demonstrative reasoning in the sciences, or all forms of rational thought. This has placed a distinctive constraint on the composition of these logical theories: their structure and internal operations have had to be totally free of any dependency on the specifics of an individual subject matter. In order to make the domain of applicability of the logic as large as possible, the techniques and methods developed in the theory have had to be uniformly applicable across all types of claims within the scope of the intended domain.

Chapter 8 explored the genesis of this principle in the logical theories of Aristotle. We saw that in his initial description of the syllogistic, Aristotle explicitly set aside reasoning which was specific to a particular subject matter (in our example, geometry). Instead, he defined the scope of logical theory to be reasoning whose structure allowed it to be "characteristic of all substance." Of course, without further argument, it was not guaranteed that there were any informational relations at all which were common to all of the interactions into which substance could enter. The critical assertion that there was such a common structure to all reasoning originated in Aristotle's metaphysics, and in particular was dictated by his account of the organization and composition of scientific knowledge. For Aristotle, the syllogistic directly captured all the different possible kinds of law-like universal relations which could connect substances and their causes, or attributes. Individual syllogisms draw their validity from the forms established by the general theory of syllogistic, and the general theory was designed to directly reflect the primitive substance-attribute relations which Aristotle hypothesized were inherent in the structure of the world. Within this metaphysical framework, then, the categorical sentence patterns which make up the syllogistic could be guaranteed to capture all the primitive propositions which constitute scientific knowledge, and the different figures and moods of the syllogistic could be guaranteed to capture all of the valid explanatory connections which could exist between categorical sentences. These two important features of Aristotle's theory, which we have described under the headings of expressive completeness and deductive completeness, were a result of the very tight integration between the theory of the syllogism and its metaphysical background. And we can see that the presence of this level of integration between the logic and the metaphysics was a fairly direct consequence of this Aristotelian logical principle of maximal scope.

Given this basic intellectual context in which logical theories were initially developed, then, we can easily see why diagrammatic representation and deductive systems for logic would not have been initially attractive. The existing diagrammatic systems in antiquity, such as those for geometry and cartography, were certainly not universal in the sense Aristotle had hoped for in his theories. Their representational methods were tied to the specific requirements of their disciplines, and their inferential techniques were not guaranteed to reflect the world's underlying metaphysical structure. Further, because of this close linkage between Aristotle's logic and his metaphysics, as long as Aristotle's basic views persisted, there was very little motivation for any sort of radically new formulation of logic, as a diagrammatic system would have undoubtedly been perceived. Syllogistic theory itself appeared to be perfect and complete, thoroughly adequate to its defined task, and admitting only of minor tinkering, such as the additional distinctions of the forms and moods developed during the medieval period. Kant himself articulated the received view on Aristotle's approach, writing in the late eighteenth century that:

> There are but a few sciences that can come into a permanent state, which admits of no further alteration. To these belong Logic and Metaphysics. Aristotle has omitted no essential point of the understanding.[1]

Further, if the theory of the syllogism was accepted as essentially complete, as Kant's words certainly show it to be, then any competing account of logic

[1] Kant, *Introduction to Logic*, quoted in Haack 1974, pg. 27.

would either be defective in some way, or be a mere notational variant of well-established syllogistic theory.

Euler's eighteenth century circle-oriented formalism was the first attempt at a diagrammatically-based logical system which had the potential to challenge the intellectual dominance of syllogistic theory. As we have described, the medieval logicians had previously created several schema which graphically depicted the basic relations of the syllogistic, but all of these lacked a deductive methodology, and they were never presented as rival logical systems to the syllogistic. They were simply viewed as cognitively appealing teaching tools and mnemonics by which the classical syllogistic forms could be made clear. Euler's system, however, was fundamentally different from these earlier attempts in two ways. First, he provided a graphical inference system which had roughly the same power as Aristotle's syllogistic figures. Second, and more significantly, the workings of the representational and inferential structures of his circles depended on going significantly beyond the Aristotelian dogma upon which logic had implicitly depended for the prior two millennia. For purposes of representation and reasoning, Euler understood categorical sentences as claims about the interaction between classes of objects, instead of as claims about primitive predication relations between substances and attributes. This was, *prima facie*, incompatible with Aristotle's carefully constructed harmony between his logical theory and his underlying metaphysics. Without a further account of the linkage between truths about extensions which Euler's system directly modeled, and the general relation of validity which Euler was trying to capture, Euler's system could not claim the philosophical pedigree of the classical syllogistic, and so represented a new direction in logic. Euler himself clearly did not see his circles as involving anything so radical, and he proceeded to structure his system around reasoning with the traditional syllogistic forms, and (like the medieval logicians) justified his graphical system based on its ease of apprehension. However, even though Euler circles were philosophically problematic, and their use was largely confined to elementary logic texts, his system was a major turning point in our investigation. Through his use of extensionally-based interpretations as a foundation for logic, Euler provided the first diagrammatic system of logic which could be claimed to have the same kind of expressive and deductive completeness as Aristotle's syllogistic.

The use of the extensional interpretation of predicates in reasoning, which was reasonably novel at the time Euler constructed his system, was at the center of Boole's initial work in class-based symbolic logic roughly a century later. Though Boole's formulation of his theory of logic did not use diagrams, his work is significant to us because it provided the foundation for an influential new view of the way that logical theories should be formulated, which was related more to equational reasoning in mathematics than to the classical syllogistic forms. More philosophically important, however, is the fact that Boole's work in logic was also explicitly founded on a new and non-Aristotelian metaphysical theory: the Kantian-inspired idea that logical theories draw their validity from their ability to model the workings of a set of abstract and universal laws of rational thought. This substitution of one logical background for another allowed Boole to assert that the scope of his proposed logical theory was indeed universal – ranging over that fragment of reasoning which is common to all rational thought – and so implicitly guaranteed that his theory would conform to the principle of maximal scope. For our purposes, Boole's logical theory and background metaphysics also provided a much more natural framework in which diagrammatic logical systems could be created. Venn's diagram system was explicitly rooted in Boole's logical approach, and involved transferring Euler's basic idea of representing the concept extensions by areas into a Boolean framework, and ridding Euler's system of its architectural dependence on the syllogistic. Venn diagrams were the first true diagrammatic systems for symbolic logic, and even though they were both expressively and deductively too weak to have the broad scope required of an orthodox logical theory, they served as the inspiration for a much more adequate system: Peirce's system of existential graphs.

At this point, we should observe that our discussion of the diagram systems of Euler, Venn, and Peirce has served to reinforce our earlier position about the possibility of diagrammatic systems in syllogistic logic. From our examination of these three systems, we have found additional strong evidence that the metaphysical background of logic – the perceived properties of the large-scale external domains whose common informational relationships were the target of logical theories – has always had a dominant impact on the structure of these logical systems and hence, on the possibility of diagrams in logic. For Aristotelian-based theories, I have shown that this background has driven the choice of the particular subject-predicate structure of the sentential formulation of the syllogism. For theories based in Boole's class-logic interpretation of the laws of thought, I have shown that it has warranted the diagrammatic systems of Venn and (antecedently) of Euler, and at the same time explained why these theories were not widely accepted. And, for Peirce's account of logic, I have shown how his background philosophical assumptions licensed a unique and central role for diagrammatic systems within the context of the other theories into which his logic must fit. This evidence is sufficient to support my claim that it is not due to a lack of creativity that diagrammatic systems were not developed before the time of Euler, Venn, or Peirce. Rather, it is because the representational structure and deductive mechanisms of any logical theory are tied, to a large but often unacknowledged extent, to the metaphysical and epistemological context in which it is developed. Certainly, the creation of robust diagrammatic systems for logic has always been hampered by the fact that this context has always involved a commitment to the principle of maximal scope; and the expressive capabilities required by this commitment have often guided the theoretical formalisms toward sententially-based systems. But, it is a central contention of this book that the primary reason why, for example, Aristotle did not develop Venn diagrams, is that such a logical system *would not have made sense* within the metaphysical environment of the time.

This basic analysis was confirmed in section 10.5, which examined the logical work of Frege and Hilbert, and their joint development of the framework under which research in logic is performed today. Recall that we argued that Frege's strong opposition to allowing into mathematics or logic any sort of proof technique involving spatial intuition would make problematic the use of any kind of non-trivial diagrammatic logical systems. His own *Begriffsschrift* notation, which consisted of a two-dimensional graph of terms and formulae connected by different kinds of lines, was apparently the limit of Frege's tolerance in this respect; he would certainly have rejected diagram systems which employed regions or sophisticated geometric relations to denote meaningful relations as too dependent on the workings of this spatial intuition. However, this rejection of graphically-based methods in proof was grounded in the Boolean idea that logical principles should be immediately reflective of an underlying set of universal laws of thought, and that these laws, in order to have the correct degree of generality, must be independent of particular experience and knowable completely *a priori*. Following Kant's arguments, Frege noted that all of our knowledge of actual spatial relations must depend on our ability to apply motion to the contents of the imagination, and to intuitively discover the results of such motion, and hence he concluded that the ability to perform geometric reasoning could be seen to be essentially synthetic and based in experience. Here, again, we can see the critical role played by assumptions about the character of the metaphysical background against which logical theories acquire their validity.

Finally, with Hilbert's work on logical theory, we saw a significant reinterpretation of the principle of maximal scope. Although Hilbert's reference-free view of the structure of logical theories guaranteed that they could be applied without regard to any particular subject matter, and thus in a sense were universal, he was also careful not to limit the domain of acceptable logical theories to the modeling of large-scale portions of general reasoning, such as the common fragment of scientific reasoning which Aristotle targeted, or the sorts of inferences which are shared by all rational thought. In his philosophy, logical theories could be freed from most of the traditional requirements on their expressive power or deductive capability, as long as they did not prejudge the sorts of domains to which they could be applied. And, in place of the classical standard that logical theories should be assessed by their ability to model these large-scale fragments of reasoning, Hilbert proposed several internal requirements on logical systems, most notably that the proofs which they generated would be checkable using only the methods available in the *finite Einstellung*. These requirements, for Hilbert, had the effect of foreclosing the possibility of diagrammatic proof systems in logic whose representational techniques went beyond the relatively simple devices used in the *Begriffsschrift*.

Therefore, this part of the book shows us how the changing perceptions of the metaphysical context and acceptable subject matter of logical theories governed the acceptability of diagrammatic logical systems. Parallel to the conclusions of our study of geometry in Part I, these systems could only be developed when the representational and deductive structures which they offered could be adequately matched to the ontological properties of the underlying domain of interpretation. Throughout the history of logic which we have examined, the choice of this domain has been shaped by certain principles which are external to the theories, but part of the disciplinary definition of logic itself. These principles, such as the Aristotelian principle of maximal scope and the Fregean prohibition against intuition-based proof methods, account for the overall absence of diagrammatic systems from the mainstream of logic. They also show us, as well as possible, that there are good explanations which go beyond lack of creativity which account for why powerful diagrammatically-based systems of logic were not developed by the classical or medieval logicians, and account for why the systems which were developed in the nineteenth century had the character that they did.

12

Conclusion

I began this book with the following broad goal: to explain why diagrammatically-based expressive and deductive methods are not among those which are formally recognized by the current theories of reasoning in geometry and logic. The primary claim of the book was that the underlying reason for the rejection of diagrams was related to the philosophical background of each discipline. More specifically, I have argued that the decision to reject diagrams in these disciplines has always been fundamentally driven by the prevailing epistemic and ontological views about the discipline's subject matter. I have made this argument by closely examining the evolution and major intellectual turning points of both logic and geometry, and emphasizing three key themes which were common to both histories: the changing attitudes toward the role of intuition in the procedures and formalisms of formal proof; the historical contrast between the very general subject matter which Aristotle defined for logic and the specific subject matter which has traditionally characterized geometry; and the ways in which both logic and geometry were affected by the drive for foundational rigor which characterized mathematics during the nineteenth century. Let me conclude briefly reviewing the role of these themes in the overall argument, and by making some closing observations about this goal and the possible future directions for this work.

One of the most striking things which our investigation has demonstrated is the *complimentary* nature of logic and geometry with respect to the evolution of the core modern notions of axiomatic systems, and in particular with respect to their rejection of diagrammatic techniques. The development of attitudes governing the use of intuition as a proof method in geometry, and the eventual complete reversal of these attitudes and adoption of the corresponding ones in logic, is an important example of this. Recall that in Euclid's proof system, assertions about equality between figures could be justified by appeal to superposition. Despite the misgivings Euclid apparently had concerning the use of this method, superposition remained a critical component of his proof system, and the fundamental requirement that the interactions of geometric components be visualizable in intuition at every step of the proof became an important feature in linking Euclidean proof systems to the domain about which they were reasoning. Descartes' analytic geometry represented a significant step away from this ontological dependence on visualizable figures, and Poncelet's introduction of ideal points and the ideal line in projective geometry further moved geometry and geometric proof still further away from the domain of the strictly visualizable. With the invention of consistent non-Euclidean geometries by Loebachevsky and Bolyai, it finally became clear that deduction techniques which were based on spatial intuition were not adequate to the demands of formal proofs in the new geometries, and diagrammatic methods were eventually banished from geometric proof in the work of Pasch. Importantly, however, this distrust of intuition in geometry was mirrored by a parallel distrust of the use of intuition in logical calculi, arising from the rigorization and sharpening of the proof methods in the calculus of algebra and analysis. We discussed evidence of this in Frege's hostility toward intuition in his work on a deductive system to underlie arithmetic. Thus, by the end of Part II of this book, we were in a position to see that Hilbert's development of a common, unified position forbidding the use of diagrammatic methods in axiomatic systems, and indeed much of his theory of the function of the *finite Einstellung* in proofs, has its roots in the traditions of both logic and geometry.

The complementary aspects of logic and geometry are also now clear with respect to the role of their differing subject matters in the development of the modern theory of axiomatic systems. In our examination of the work of Descartes, Poncelet, Pasch, Euler, Venn, Frege, and Hilbert, we have shown how the major events in the development of diagrammatic systems in logic and geometry were linked to changes in the perceptions of the underlying modeled domain. Logic, as we have observed, was initially conceived as a discipline which was concerned with just those patterns of reasoning which were completely general and appropriate to any sort of Aristotelian substance. The early guarantees of validity for the syllogistic figures were founded on an Aristotelian ontology of properties, and on a metaphysics in which the possible relationships between scientific truths of the world were directly reflected in the structure of the syllogistic sentences. Geometry, on the other hand, was classically rooted in reasoning about a very specific subject matter – the extensional properties of objects in Euclidean space – and the guarantees of the truth of geometric theorems were connected to intuitions about these properties. This meant that the traditional diagrammatic methods of geometry, and indeed any sort of domain-specific reasoning procedures, were viewed as outside of the scope of logic proper. This universalist restriction on the subject matter of logic remained unchanged even as the classical metaphysics upon which it was founded was superseded by Kantianism, and (in turn) by other forms of nineteenth-century metaphysics. When the invention of non-Euclidean geometry forced geometers such as Pasch to reconstruct their proof systems and underlying philosophical assumptions to be neutral between several different geometries, diagrammatic methods were argued to be illegitimately tied to a particular Euclidean interpretation of the terms. Furthermore, by the time of Hilbert, the Aristotelian/Kantian doctrine of *logica magna* was starting to give way to an attitude of *logica utens*, which (in a way similar to the split between pure and applied geometry) promised relative freedom from metaphysical entanglements in return for accepting an essential indeterminacy of reference in the objects of the logic. So, by tracing the evolution of attitudes towards the subject matter of logic and geometry, we can understand the influence of this factor in Hilbert's choice to exclude diagrammatic methods when merging of the foundations of logic and of geometry into a single unified theory of axiomatic systems. This restriction was a natural outgrowth of the point to which logic and geometry had both developed with respect to their views on their subject matter.[1]

Finally, we have described in this book how the overall trends in nineteenth-century mathematics affected both logic and geometry, and helped to bring their foundations together. For most of their histories, the stories of logic and geometry were separate and nonoverlapping, and it was only in the late eighteenth-century work of Kant that their philosophical foundations were first treated together. However, we have now seen how the mathematical ferment of the nineteenth century, especially the general concern with rigor and the development of the foundations of analysis and modern algebra, helped create and foster a situation in which projective and non-

[1] This point in particular is an interesting one to speculate upon. At the point at which the proof systems and theoretical background for the new logics were being developed, we have shown that there were no examples of serious diagram use in exact reasoning. The deductive systems of geometry – the traditional paradigm examples of axiomatic systems – had been developed to a stage at which they no longer depended on diagrams, and indeed actively discouraged their use. Because of this, the rich diagrammatic tradition in geometry did not influence the development of representations for modern axiomatic systems. However, it is interesting to wonder what might have happened if the inventions of symbolic logic and non-Euclidean geometry had been separated by a century, or had happened prior to the nineteenth-century revolution in mathematics. Would powerful diagrammatically-based deductive systems been developed and accepted into logic if the historical ordering had been different?

Euclidean geometries could be developed, and in which symbolic logic could be conceived. During this time, it first became possible to attempt to model the calculus of logic after the calculus of algebra, and in this way give substance to Leibniz's conception of an *ars combinatoria*. Also, the increasing acceptance of abstraction in nineteenth-century mathematics helped to legitimize both Pasch's call to divide geometry into pure and applied subfields, and Hilbert's attempts at promoting an attitude of *logica utens* in his theory of axiomatic systems. So, our focus on the mathematics of this period has shown how the current marginalization of diagrammatic techniques is closely related to the overall changes in mathematical philosophy and practice which occurred in the mid-nineteenth century.

Each of these three themes of intuition, subject matter, and mathematical context has helped to bolster our main contention: that the possibility of diagrammatic methods in formal proofs in logic and geometry has always been primarily dependent on the characteristics of the metaphysical and ontological theories under which they are carried out. Together they constitute a compelling argument. In conclusion, however, we note that by affirming the overall claim with which we started, we have not at the same time provided a decisive, knockdown argument against the possibility of sophisticated logical or geometric theories which incorporate diagrammatic systems of reasoning. Instead, what we have shown is that the current attitudes about the legitimate representational formalisms in modern axiomatic systems result from a confluence of intellectual and historical currents in the late nineteenth century. However, the modern development of symbolic logic, with its essential abstraction from subject matter and its clear distinction between the syntactic and the semantic, has provided a new conceptual apparatus and philosophical background, which may be more congenial to the development of formal systems of diagrammatic reasoning. Through the use of these theoretical tools, it may be possible to construct diagrammatic deductive systems and plausibly argue that they stand on equal footing with traditional sentential ones.

Let me close by speculating a bit more on this point. Given all of the historical arguments we have explored against the use of diagram systems in formal proof, what lessons can we draw for current work on sophisticated diagrammatic systems for logic? The answer, it now appears, is intimately tied up with the larger philosophical perspective that one adopts toward logic and mathematics. If one accepts a Peircean view of logic, for example, then the development of strong diagrammatic systems of deduction is certainly licensed, and indeed is even encouraged. However, accepting Peirce's philosophy may also (depending on the individual) involve some less desirable consequences, such as giving up certain historically-endorsed notions of necessary truth, adopting a fairly cumbersome theory of language and semiotics, and encouraging a shift towards the philosophical stance of pragmatism. If instead one adopts Boole's and Frege's attitudes toward logic – that logical theories should be directly modeled on a set of universal laws of thought – then their arguments about the necessity of restricting the mechanisms of logic to those which were purely analytic and *a priori* come into play. For Frege, logic's field of study was the most general laws of concept combination, and therefore logic's content and techniques must be completely free of synthetic or *a posteriori* influences, including all traces of intuition or other psychological notions. Several philosophers, including Kant and Husserl, have given powerful arguments to the effect that our knowledge of space must be synthetic, and these arguments would have to be refuted in order to justify adding diagrammatic techniques to a Fregean philosophy of logic.

The most interesting possibility arises if one wishes to develop and employ diagrammatic logical systems for mathematics, and simultaneously maintain a Hilbert-style formalist philosophy of mathematics. As we have seen, Hilbert's solution to the traditional metaphysical concerns of the philosophy of mathematics was to exchange traditionally difficult questions about the external ontology with purely intrasystem questions about the completeness, consistency, and finiteness of the foundational logic and axiom set. In order to incorporate diagrammatic methods into this view of logic, it would have to be argued that the diagrams could be coherently assumed to be reference-free, and that any proofs which involved diagrams should be analyzable via the mechanisms of the *finite Einstellung*. This would naturally lead into the accompanying issues about the role of intuition in proofs of this type, and the place of diagrammatic methods in this framework. While his remarks concerning the use of diagrams in geometry make clear that Hilbert himself would have rejected any such attempt to legitimize diagrammatic logical systems, some recent work on computer-aided environments for working with such systems seems to suggest that reasoning with diagrams can be carried out without the need to appeal to an intrinsic content or interpretation of these diagrams.[2] At any rate, the possibility of evaluating the strength of Hilbert's or Frege's arguments in the context of current research in diagrammatic systems in logic is a promising topic, and one which awaits further research.

---

[2] See, *e.g.*, Allwein and Barwise 1996 for descriptions of this work.

References

Allwein, Gerard, and Jon Barwise (eds.). 1996. *Logical Reasoning with Diagrams*. New York: Oxford University Press.

Anglin, William. 1994. *Mathematics: A Concise History and Philosophy*. New York: Springer-Verlag.

Aristotle. *A New Aristotle Reader*. ed. 1987 by J. L. Ackrill.. Princeton: Princeton University Press.

Arnauld, Antoine. 1662. *La Logique, Ou L'art de Penser*. (trans. 1964 by James Dickoff and Patricia James as *The Art of Thinking*). Indianapolis: Bobbs-Merrill.

Bacon, Roger. 1928. *The Opus Majus of Roger Bacon*. (trans. 1928 by Robert Burke). Philadelphia: University of Pennsylvania Press.

Barwise, Jon. 1986. Logic and Information. *The Situation in Logic*. (ed. 1989 by Jon Barwise). Stanford: CSLI Publications.

Barwise, Jon, and John Etchemendy. 1991. Visual Information and Valid Reasoning. In Allwein and Barwise 1996.

Barwise, Jon, and John Etchemendy. 1994. *Hyperproof*. Stanford: CSLI Publications.

Boole, George. 1854. *An Investigation of the Laws of Thought*. London: Macmillan; reprinted by Dover, New York, 1958.

Cayley, Arthur. 1898. *The Collected Mathematical Papers of Arthur Cayley*. Cambridge: At the University Press.

De Morgan, Augustus. 1864. On the Syllogism IV and the Logic of Relations. *Cambridge Philosophical Transactions*. v. 10.

Descartes, Rene. 1637. *The Geometry of Rene Descartes*. (trans. 1954 by David Smith and Marcia Latham). New York: Dover.

Euler, Leonhard. 1846. *Letters of Euler on Different Subjects in Natural Philosophy, Addressed to a German Princess, With Notes and a Life of Euler.* Vol. 1. (trans. 1846 by David Brewster). New York: Harper and Brothers.

Eves, Howard, and Carroll Newsom. 1965. *An Introduction to the Foundations and Fundamental Concepts of Mathematics.* New York: Holt, Rinehard and Winston.

Fishback, William. 1962. *Projective and Euclidean Geometry.* New York: Wiley.

Frege, Gottlob. 1879. Begriffsschrift, A Formula Language, Modeled Upon That of Arithmetic, For Pure Thought. In van Heijenoort 1967.

Frege, Gottlob. 1884. *The Foundations of Arithmetic.* (trans. 1950 by J. L. Austin). 5th ed. Oxford: Blackwell.

Frege, Gottlob. 1980. *Philosophical and Mathematical Correspondence.* (ed. 1980 by Brian McGuinness, trans. by Hans Kaal). Oxford: Blackwell.

Frege, Gottlob. 1984. Methods of Calculation Based on an Extension of the Concept of Quantity. *Collected Papers on Mathematics, Logic, and Philosophy.* (ed. 1984 by Brian McGuiness, trans. by Hans Kaal). Oxford: Blackwell.

Friedman, Michael. 1997. Geometry, Construction, and Intuition in Kant and his Successors. *Between Logic and Intuition: Essays in Honor of Charles Parsons* (ed. by G. Scher and R. Tieszen), forthcoming.

Gardner, Martin. 1958. *Logic Machines and Diagrams.* New York: McGraw-Hill.

Haack, Susan. 1974. *Deviant Logic: Some Philosophical Issues.* New York: Cambridge University Press.

Hallett, Michael. 1994. Hilbert's Axiomatic Method and the Laws of Thought. *Mathematics and Mind.* (ed. 1994 by Alexander George). New York: Oxford University Press.

Hammer, Eric. 1995. *Logic and Visual Information.* Stanford: CSLI Publications.

Heath, Sir Thomas. 1908. *The Thirteen Books of Euclid's Elements*, 2nd ed. (3 vols.) Cambridge: Cambridge University Press; reprinted by Dover, New York, 1956.

Heijenoort, Jean van. 1967. *From Frege to Gödel: A Source Book in Mathematical Logic* (ed. 1967 by Jean van Heijenoort). Cambridge MA: Harvard University Press.

Heijenoort, Jean van. 1985. Absolutism and Relativism in Logic, in *Selected Essays.* Napoli: Bibliopolis.

Hilbert, David. 1899. *The Foundations of Geometry.* (trans. 1971 by Leo Unger from the 10th German ed., rev. and enl. by Paul Bernays). La Salle: Open Court.

Hilbert, David. 1900. Mathematical Problems. *Readings in the Philosophy of Mathematics.* (ed. 1994 by William Ewald). Oxford: Clarendon Press.

Hilbert, David. 1904. On the Foundations of Logic and Arithmetic. In van Heijenoort 1967.

Kant, Immanuel. 1787. *The Critique of Pure Reason.* (trans. 1929 by Norman Kemp Smith). New York: St. Martin's Press.

Kline, Morris. 1972. *Mathematical Thought from Ancient to Modern Times.* Oxford: Oxford University Press.

Kneale, William, and Martha Kneale. 1962. *The Development of Logic.* Oxford: Clarendon Press.

Knorr, Wilbur. 1975. *The Evolution of the Euclidean Elements: A Study of the Theory of Incommensurable Magnitudes and Its Significance for Early Greek Geometry.* Dordrecht: Reidel.

Knorr, Wilbur. 1986. *The Ancient Tradition of Geometric Problems.* Boston: Birkhauser.

Manders, Ken. 1994. Diagrammatic Contents and Representational Granularity. unpub. manuscript.

McKeon, Richard. 1972. *Introduction to Aristotle.* 2nd ed. Chicago: University of Chicago Press.

Mueller, Ian. 1981. *Philosophy of Mathematics and Deductive Structure in Euclid's Elements.* Cambridge, MA: MIT Press.

Nagel, Ernest. 1939. The Formation of Modern Conceptions of Formal Logic in the Development of Geometry. *Osiris* v.7.

Plato. *The Collected Dialogues of Plato.* (ed. 1961 by Edith Hamilton and Huntington Cairns) Princeton: Princeton University Press.

Peirce, Charles. 1931. *Collected Papers.* 8 vols. (ed. 1931 by Charles Hartshorne and Paul Weiss). Cambridge MA: Harvard University Press.

Peirce, Charles. 1940. Logic as Semiotic: The Theory of Signs. *Philosophical Writings of Peirce.* (ed. 1940 by Justus Buchler). London: Routledge; reprinted by Dover, New York, 1955.

Polythress, V., and H. Sun. 1972. A Method to Construct Convex Connected Venn Diagrams for Any Finite Number of Sets. *Pentagon* 31.

Poncelet, Jean Victor. 1865. *Traité des Propriétés de Figures.* Vol. 1. Paris: Gauthier-Villars.

Reed, David. 1995. *Figures of Thought: Mathematics and Mathematical Texts.* London: Routledge.

Salmon, Merrilee. 1989. *Introduction to Logic and Critical Thinking.* 2nd ed. New York: Harcourt Brace Jovanovich.

Shin, Sun-Joo. 1994. *The Logical Status of Diagrams.* New York: Cambridge University Press.

Sowa, John. 1984. *Conceptual Structures: Information Processing in Mind and Machine*. Reading, MA: Addison-Wesley,

Tennant, Neil. 1986. The Withering Away of Formal Semantics? *Mind and Language*, v. 1, no. 4.

Venn, John. 1884. *Symbolic Logic*. (reprinted 1972.) New York: Chelsea.

Index